AFTERGLOW OF CREATION

From the Fireball to the Discovery of Cosmic Ripples

AFTERGLOW OF CREATION

From the Fireball to the Discovery of Cosmic Ripples

Marcus Chown

University Science Books
Sausalito, California

University Science Books
55D Gate Five Road
Sausalito, CA 94965
Fax: (415) 332-5393

Designer: *Robert Ishi*
Illustrators: *John and Judy Waller*
Compositor: *Wilsted & Taylor*
Printer and binder: *Maple-Vail Book Manufacturing Group*

This book is printed on acid-free paper.

Library of Congress Cataloging-in-Publication Data

Chown, Marcus.
Afterglow of creation / Marcus Chown.
p. cm.
Includes index.
ISBN 0-935702-40-7 (hc)
1. Cosmic ripples. 2. Cosmic background radiation. I. Title.
QB991.C64C487 1995
523.18—dc20
95-38781
CIP

Printed in the United States of America
10 9 8 7 6 5 4 3 2 1

To my mum and dad...My mum for talking to me incessantly from the moment I was born. And my dad for giving me Dr. H.C. King's *Book of Astronomy* as a Christmas present and getting me out of bed to see the Moon landings on television.

No one's mum and dad could have done more.

—MC

CONTENTS

PREFACE

When *Afterglow of Creation* was published in Britain, the response outstripped all my expectations. I received many letters from people who had never read a science book before. Piano tuners telephoned me at work. Old ladies sent me poems inspired by the book. School children wanted to know how they could become astronomers. In a national newspaper, the international violinist, Yehudi Menuhin named *Afterglow* his favorite book. The BBC bought the television rights. I was even interviewed on a breakfast television program by two puppets from the planet Zog!

And to top it all, a popular science magazine purchased 200,000 copies to give away to its readers, making *Afterglow* the best-read (if not best-selling) science book since Stephen Hawking's *A Brief History of Time.* The book was even runner-up for a prestigious science book prize (oh well, you can't win 'em all!)

But although I was delighted with the response, I could not hide my disappointment that *Afterglow* was not published in the US. This was particularly frustrating because the book is about American science, an American space project, and I myself even spent time at an American university! It is therefore a great pleasure to finally see a US edition of the book. This would not have happened without Bruce Armbruster and University Science Books. I am indebted to them both.

Marcus Chown

ACKNOWLEDGMENTS

Many people have helped in the writing of this book. In particular, I would like to thank Bruce Armbruster, Jane Ellis and Neil Belton, without whom it would never have appeared.

I would also like to thank those whom I met during my research trip to the United States. All were exceptionally generous with their time. In particular, I would like to thank David Wilkinson, Jim Peebles and Bob Dicke of Princeton University; John Mather and Charles Bennett of the Goddard Space Flight Center in Greenbelt, Maryland; Bruce Partridge of Haverford College in Philadelphia; and Robert Wilson of AT&T Bell Laboratories at Holmdel, New Jersey.

I would also like to thank Derek Martin of Queen Mary & Westfield College, London, who lent me the entire contents of his filing cabinet on the cosmic background radiation; George Smoot of the University of California in Berkeley; Robert Herman of the University of Texas at Austin; Lyman Page of Princeton University; Herb Gush of the University of British Columbia in Vancouver; John Beckman, Ken Croswell, Nigel Henbest, John Emsley, Jeff Hecht and Michael White.

I hope it goes without saying that none of these people are responsible for any errors I have made.

But most of all I would like to thank my wife, Karen, who put up with me getting up to write each day at the crack of dawn and whose comments gave me confidence that my explanations were not too opaque. I hereby promise to do the hoovering, ironing and washing for the next six months, and to walk our deaf, blind, and arthritic dog, Daisy.

PHOTO CREDITS

University Science Books gratefully acknowledges the following individuals, photo archives, and organizations who supplied the printed images herein.

Chapter One: p. 10. Hubble on Mount Wilson — Photo by M. Harwood and courtesy of the AIP Emilio Segrè Visual Archives; p. 11. Andromeda — courtesy of Palomar Observatory California Institute of Technology; p. 15. Vesto Slipher — courtesy of AIP Emilio Segrè Visual Archives; p. 17. From *The First Three Minutes* by Steven Weinberg, p. 29, copyright 1977, 1988 by S. Weinberg. Reprinted by permission of BasicBooks, a division of HarperCollins Publishers, Inc.; p. 19. From *The Physical Universe* by Frank H. Shu, pg. 352. Copyright 1982 by University Science Books. Reprinted by permission.

Chapter Two: p. 22. Hubble and Einstein on Mt. Wilson — courtesy of The Huntington Library, San Marino, California; p. 25. Aleksandr Aleksandrovich Friedmann — courtesy of Physics Uspekhi and AIP Emilio Segrè Visual Archives; p. 28. From *The Physical Universe* by Frank H. Shu, pg. 357. Copyright 1982 by University Science Books. Reprinted by permission; p. 33. Fred Hoyle — photograph by Philip Daly.

Chapter Three: p. 36. George Gamow — courtesy of AIP Emilio Segrè Visual Archives; p. 43. From *The Physical Universe* by Frank H. Shu, pg. 77. Copyright 1982 by University Science Books. Reprinted by permission; p. 45. Ralph A. Alpher and Robert Herman — courtesy of R. A. Alpher; p. 47. From *The Physical Universe* by Frank H. Shu, pg. 394. Copyright 1982 by University Science Books. Reprinted by permission.

Chapter Four: p. 52. Bob Dicke — courtesy of David Wilkinson; p. 56. Philip J.E. Peebles — photograph by Robert P. Matthews — courtesy of Princeton University; p. 58. Peter Roll and David Wilkinson — photograph by Robert P. Matthews.

Chapter Five: p. 66. Wilson and Penzias — courtesy of AT&T Archives; p. 68. Echo communications satellite — courtesy of NASA; p. 73. Edward A. Ohm — courtesy of AT&T Archives.

Chapter Six: p. 87. *The New York Times*. Copyright 1965 by the New York Times Company. Reprinted by permission.

Chapter Seven: p. 97. Steven Weinberg. Photograph by Louise Weinberg; p. 99. Walter Adams — courtesy of the AIP Emilio Segrè Visual Archives; p. 103. Bruce Partridge — courtesy of Haverford College; p. 108. Arno A. Penzias and Robert W. Wilson — courtesy of AT&T Archives.

Chapter Eight: p. 112. The Yuma experiment — courtesy of Bruce Partridge; p. 118. David Wilkinson & his team — photograph by Harold A. Wilkinson, courtesy of David Wilkinson.

Chapter Nine: p. 126. Adapted from D.P. Woody and P.L. Richards, *Phys. Rev. Lett.*, 42, 1979, 925; p. 128. Bump in the spectrum — courtesy of Paul Richards; p. 134. Universe as Swiss Cheese — after de Lapparent, Geller, and Huchra 1986, *Astrophysical Journal Letters;* graphics by M. Geller and E. Falco, copyright 1995, Smithsonian Astrophysical Observatory.

Chapter Ten: p. 140. John Mather — courtesy of NASA Goddard Space Flight Center; p. 145. The Space Shuttle Challenger — courtesy of NASA: p. 147. Charles L. Bennett — courtesy of Charles Bennett; p. 153. Launch of NASA's COBE — courtesy of NASA.

Chapter Eleven: p. 156. Adapted from *Discovering the Cosmos,* by Robert Bless, copyright 1996 by University Science Books; p. 162. Herb Gush — courtesy of Herb Gush.

Chapter Twelve: p. 171. *The New York Times*. Copyright 1995 by the New York Times Company. Reprinted by permission.

Chapter Thirteen: p. 179. George Smoot — courtesy of Lawrence Berkeley Laboratory.

Chapter Fourteen: p. 188. Fritz Zwicky. Courtesy of the AIP Emilio Segrè Visual Archives; p. 194. Alan Guth — courtesy of the MIT News Office; p. 197. A short movie of the Universe — courtesy of John Mather.

Chapter Fifteen: p. 204. Photo by Ken Ganga, courtesy of Lyman Page; p. 209. Keck telescope.

Color Plate 1, COBE image — courtesy of NASA.
Color Plate 2, COBE image — courtesy of NASA.

AFTERGLOW OF CREATION

From the Fireball to the Discovery of Cosmic Ripples

INTRODUCTION

The world through microwave eyes

It's a crystal-clear night far away from the bright lights of a big city. A luminous full Moon is pulling itself free of the treetops. Against the velvet-black sky stars are winking like diamonds. But the night sky is not all it seems . . .

The visible light our eyes see makes up only a vanishingly small portion of all the light that is streaming through the Universe. Raining down on the Earth from space is a ceaseless torrent of invisible "light."

For most of human history we have been entirely blind to this light. But in recent years astronomers have opened up our eyes. New telescopes have been built which can see x-rays, infrared light, radio waves and every other kind of invisible light. Now, for the first time, we can behold the greater glories of the Universe.

Imagine that you can see what the astronomers see simply by putting on a pair of "magic" glasses. To "tune" them to different types of light you need only twiddle a knob on the frame. No longer are you almost blind. Now you can have infrared eyes, radio eyes, eyes that see ultraviolet light, gamma rays or x-rays.[1]

What could you see through these impressively enhanced lenses?

At first, nothing appears to be changing. Then you realize that the Moon is fading. So, too, are most of the stars. Soon the Moon is hardly visible and the stars have begun to wink out one by one. But

[1]Strictly speaking, you would have to go into space to use your magic glasses because most invisible light is absorbed by the air in the atmosphere. But don't let that worry you. This is only a story!

as the stars disappear new ones pop into view in places where no stars were visible before. Some of the new stars are shrouded in clouds of misty white.

This is the ultraviolet sky. Your glasses are registering the kind of invisible light that causes sunburn when you lie too long on a beach. Only the very hottest stars shine brightly with ultraviolet light.

Twiddle on.

The stars change again. Now there are no familiar signposts in the heavens. The intensely bright pinpricks that dot the sky mark places where stars are cannibalizing other stars and where blisteringly hot gas is plunging headlong into black holes. Wherever matter is heated to millions of degrees it shines brightly with x-rays.

Keep twiddling.

Everything is fading now. We have come to gamma rays, the most energetic light in the Universe, created by the most violent events imaginable. Now the sky looks utterly black.

But there is a tiny brilliant flash of light. You turn your head to stare. But there is nothing to see. The black sky is utterly empty. But if you were very patient indeed and watched the gamma ray sky for several days at a stretch you would see another brilliant flash from an entirely different part of the sky. And after a few more days you would see another. Nobody knows what these flashes are. Astronomers have called them "gamma ray bursters" and they are one of the great unsolved mysteries of the Universe.

There isn't anything more to see by tuning any further—except darkness and yet more darkness. Turn the knob back the other way, through x-ray and ultraviolet skies to the visible sky with its full Moon and familiar stars. But don't stop. Keep going. Keep turning.

You are now seeing infrared light. Instead of the Universe's hot bodies, you are seeing relatively cool ones. Even human beings give out infrared light. It's the same kind of light earthquake rescue teams use to detect people trapped beneath rubble.

The Moon has reappeared in the sky. But instead of shining brightly from reflected sunlight, it is glowing dully from its own meagre internal heat. The sky is full of unfamiliar stars. Cold stellar embers. There are bloated red giants in their death throes and stars so new that they are still swathed in the shimmering gas out of which they were formed.

But now you have left even the infrared sky behind. You are seeing microwaves, the same type of light used for radar and for heating food in those ubiquitous little ovens. Now if our glasses were working, something very odd would begin to happen. The sky would light up: not just part of it—all of it.

The whole sky, from horizon to horizon, is glowing a uniform pearly white. You tune further into the microwave region but the sky simply gets brighter. The whole of space seems to be glowing. It is as though you were inside a giant light bulb. And what you are seeing is quite real. It is the relic of the Big Bang, the titanic fireball in which the Universe was born. Incredibly, it still permeates every pore of space 15 billion years after the event.

There is more energy tied up in this universal "cosmic microwave background" than there is in the visible light of all the stars put together. In fact, Big Bang radiation makes up 99 per cent of all the light energy streaming through the Universe at this moment.

Yet although the technology to detect microwaves was developed for radar during the Second World War, it was not until 1965 that anyone noticed this "afterglow of creation." And even then it was noticed only by accident.

The extraordinary story of the discovery of the relic radiation from the Big Bang forms the backbone of this book. With its tortuous twists and turns, accidents and missed opportunities, it provides a wonderful example of the way in which science is really carried out.

The cosmic microwave background is the oldest "fossil" in creation. It has come to us directly from the Big Bang and has been travelling across space for 15 billion years. The cosmic microwave background was given out by matter cooling in the fireball. So it carries with it an imprint of the Universe as it was soon after the Big Bang. When you look at the microwave sky you are seeing a snapshot of the Universe 15 billion years ago.

The early Universe must have been an extremely boring place, you think. However, the beauty of this featureless, uniform Universe is that it is a lot easier for scientists to understand than a complicated one. The smoothness of the cosmic microwave background is telling us that matter in the early Universe must have also been spread amazingly smoothly throughout space. And herein lies a great puzzle. Today's Universe is anything but smooth. In fact, the Universe is full of stars and the stars are grouped together into galaxies and these galaxies in turn are linked into great chains and clusters that snake their way across space. And between these groupings of galaxies are great voids of utterly empty space. Far from being smooth, the luminous material in today's Universe has the appearance of Swiss cheese.

So how did such an uneven and complicated Universe arise from such a smooth and simple beginning?

Clearly, at some point the stuff of the Universe must have begun to clump together, like milk curdling. So although the cosmic microwave background looks remarkably smooth, it cannot be dead smooth. If we look closely at it we ought to be able to see signs of the first structures in the Universe beginning to clump under gravity soon after the Big Bang.

For more than twenty-five years, since the discovery of the cosmic microwave background, astronomers have peered at it closely. But, try as they might, they have been unable to find any variation in the brightness of the microwave background.[2] There were no signs of the lumps of matter which would later form galaxies like our own Milky Way. The evidence of the cosmic microwave background seemed to be contradicting one of our most cherished beliefs: that we and our world exist!

In 1989, NASA launched an obscure satellite called COBE (pronounced co-bee) into an orbit just above the Earth, to study the fireball radiation. This had previously been difficult because the Earth's atmosphere glows brightly with microwaves.[3]

---

[2]This is not strictly true. In the late 1970s, astronomers discovered that the microwave background is slightly hotter in the direction the Earth is moving in space and slightly colder behind us. But this is due to our motion through the microwave background and is not inherent in the background radiation itself.

[3]In fact, the ground also glows with microwaves, as do buildings, trees, people, and even clouds of hydrogen gas floating in space. These competing sources of microwaves make the uniform glow of the fireball radiation a little more difficult to spot than I have led you to believe. They explain why detecting the fireball radiation is a challenge and why it was not discovered until 1965.

COBE's sensitive instruments listened carefully for the faint whisper of the cosmic explosion that started the Universe's expansion 15 thousand million years ago. For more than two years the satellite found nothing. There were jittery mutterings among scientists.

But in April 1992, COBE hit the jackpot. It found "ripples in the cosmic background radiation." In some parts of the sky the cosmic microwave background was ever so slightly brighter than in others. It was a tiny effect. The "hotspots" in the sky were only about a part in a 100,000 hotter than the "coldspots." But the outpouring of relief among scientists was unprecedented. "It's like seeing the face of God," declared one of the scientists on the COBE team. "It's the discovery of the century, if not of all time," declared the physicist Stephen Hawking.

Many thought these remarks a little extravagant but the fact remained that COBE had found the "seeds" of galaxies in the early Universe. Those regions that were slightly denser than others would grow and grow as the Universe expanded in the aftermath of the Big Bang, getting bigger as their gravity pulled in more and more matter. They would eventually become the clusters and superclusters of galaxies we see around us today. COBE had not quite seen the face of God but it had seen the largest and oldest structures in the Universe.

At the time of the discovery the world's media went wild. The story was splashed across TV screens and the front pages of newspapers all over the planet. It is probably true that no other scientific story has ever received such blanket coverage in the media.

Why so many people lost their heads over such an obscure and esoteric story is a bizarre tale in itself and one that I tell later in this book. But before you can understand what all the fuss was about, you need to have a little cosmic background. In particular, you need to know about the Big Bang.

The story begins in the first decades of the twentieth century when a new generation of giant telescopes allowed astronomers to probe the remote depths of space and discover for the first time just what kind of Universe we were living in . . .

PART I

THE TOUGHEST MEASUREMENT IN SCIENCE

1

THE BIG BANG

How did we come to believe in such a ridiculous idea?

In December 1924, the astronomers of the world gravitated to Washington DC for the thirty-third meeting of the American Astronomical Society. It was a routine and unremarkable meeting; some of the participants had already departed to catch their trains home when, late on the last day, one man stood up in front of a half-empty auditorium, cleared his throat and began to read out a scientific paper. It had been submitted by a 35-year-old astronomer who had himself been unable to make the arduous journey east from southern California.

When the reader finished and left the podium, there must have been many in the audience who felt a sudden chill descend on the auditorium. For at long last, the human race knew the true scale of the Universe it was lost in. And it was unimaginably more vast than anyone had ever dreamed.

The absent Californian astronomer was Edwin Hubble, an ex-athlete and ex-boxer who had given up a promising career in law in order to study the heavens. In 1923, he had turned the most powerful telescope in the world—the newly built 100-inch reflector on Mount Wilson above Pasadena—onto a misty white patch in the night sky known as the Great Nebula in Andromeda. What he had made out in the outskirts of the nebula, so faint that they teetered on the very edge of invisibility, were the tiny specks of individual stars.

To understand why this changed our picture of the universe you have to realize that, at the time of Hubble's observation, most

Edwin Hubble, the man who discovered the Big Bang Universe, on Mount Wilson in California in 1923.

astronomers assumed that Andromeda was merely a cloud of glowing gas floating between the stars. But Hubble showed this was wrong. Andromeda was no nebula. It was made of stars blurred together by sheer distance. It was a vast island of stars suspended in the deep of space.

The mysterious spiral nebulae

By discovering his remote stars, Hubble had settled a fierce astronomical debate that had been raging throughout the early decades of this century. It concerned the nature of the "spiral nebulae," of which Andromeda was the largest and so most easily studied with telescopes.

The spiral nebulae had been discovered in the eighteenth century when the first generation of astronomers to use telescopes had seriously trained their instruments on the sky. Their passion was comet-hunting, so these early astronomers were irritated to discover that cluttering up the night sky were many fuzzy patches of light that could easily be confused with comets. In 1784, the French astronomer, Charles Messier, provided a valuable service to his fellow comet-hunters by publishing a catalogue of the positions of the brightest of these "vermin of the skies."

The Andromeda galaxy.

Messier's original catalogue contained one hundred and three cloud-like objects, the majority of which were spiral-shaped nebulae. At number thirty-one in the list was the Great Nebula in Andromeda. Arguably the least comet-like of all the celestial objects in Messier's list, the nebula is easily visible to the naked eye if you know where to look—a fuzzy elongated cloud about six times as big as the Moon appears in the sky. To this day, astronomers refer to it as Messier-31, or M31 for short.

The fierce debate about the nature of the spiral nebulae was inextricably bound up with the size of the Universe for the following reason. If the spiral nebulae were clouds of glowing gas, as most astronomers maintained, then they must be near the Earth. Glowing gas simply did not shine brightly enough to be visible at great distances.

But others argued that the spiral nebulae were great islands at enormous distances from the Earth. They only appeared like clouds of glowing gas because distance had blurred their stars together.

At the time, it was known that our Sun belonged to a large stellar swarm called the Milky Way. The Milky Way is a flattened roundish formation of stars similar in appearance to a compact disc. In the night sky it appears as a misty band stretching across the heavens, but that is only because we see it edge-on from our position inside it.

In the early part of this century, many astronomers believed that the Milky Way was the entire Universe and that nothing existed beyond its limits. If the spiral nebulae were shown to be beyond the Milky Way then this idea of a visible Universe would be blown apart.

The moment Hubble found stars in Andromeda, it began to look as though it was indeed beyond the Milky Way. But unless he could discover its exact distance, Hubble could not tell for sure.

Fortunately, among the Andromedan stars Hubble was able to identify were very unusual stars known as Cepheids. And these enabled him to settle the question once and for all.

To an astronomer, finding Cepheids is like scouring a vast expanse of beach and stumbling on a handful of jewels sparkling in the sand. The reason is that it is always possible to determine the distance to a Cepheid, which is usually impossible with ordinary stars. If you see two stars and one appears brighter than the other, it is impossible to tell whether the bright one is really brighter or whether it is simply closer. But there is a way of telling how bright Cepheids really are.[1] So if an astronomer sees two similar Cepheids and one is brighter than the other, he can be certain that the bright one really is closer.

Building blocks of the Universe

Hubble compared the Cepheids he had found in Andromeda with those in the Milky Way. He found that the Cepheids in Andromeda were immensely further away. Andromeda was at a truly enormous distance. It was a "galaxy," a vast island of billions of stars floating in space far beyond the limits of the Milky Way.

[1]Cepheids are "variable" stars that brighten and dim periodically. In 1908, Henrietta Leavitt discovered that how long they take to do this is related to how bright they really are. So to know a Cepheid's true brightness, you need only measure the "period" of its light variation.

How did we come to believe in such a ridiculous idea?

If Andromeda was a separate galaxy, then the implication of this was obvious to Hubble. The Milky Way must be a galaxy as well. Though it looked like a flattened disc of stars from our vantage point, it, too, was a spiral galaxy, a giant fiery pinwheel turning ponderously in space.

And if Andromeda was a galaxy, all the other spiral nebulae littering the heavens must also be galaxies, giant beacons of stars burning brightly out of the black depths of space. Far from being all of creation, the Milky Way was merely one galaxy among countless billions of others scattered throughout space. Galaxies like Andromeda, which appeared large and bright in our sky, were simply close neighbors of the Milky Way. The small and faint galaxies were at enormous distances.

Hubble had demonstrated just how large our Universe really is. He had identified the building blocks of the Universe—immense pinwheels and spheroids of stars. They crowded space all the way out to the very limits probed by the largest telescopes, dwindling finally to mere specks of light (see Color Plate 1).

Today, the Universe we see with our telescopes is about 20 thousand million million million miles across. If that gives you a headache, try imagining the Universe as a sphere just a mile in radius. In this shrunken Universe, our Galaxy,[2] the Milky Way, which has more than 100 thousand million stars, floats at the center and is roughly the size and shape of an aspirin.

But the Milky Way is not alone in space. Galaxies tend to congregate in "clusters" and our Milky Way is no exception. It belongs to a meager cluster of galaxies called the Local Group. Of the cluster's couple of dozen galaxies, only one—the Andromeda galaxy—is sizeable. Andromeda is another aspirin floating in space a little over four inches away.

The nearest large cluster of galaxies to our own is the Virgo Cluster, which contains about two hundred galaxies. In this Lilliputian universe, the galaxies of the Virgo Cluster occupy the volume of a basketball and are about three yards away.

Some other more distant clusters may contain many thousands of aspirin-sized galaxies and these clusters may be many

[2]Astronomers always give our Galaxy a capital "G" to distinguish it from other galaxies.

yards across. And clusters of galaxies in turn form clusters, which astronomers call "superclusters." Aspirin-sized galaxies crowd space out to the edge of observable Universe a mile away.

The fleeing nebulae

Hubble had succeeded in identifying the major constituents of the Universe—the galaxies—and provided some sense of the vastness of the cosmos that they inhabited. But he had yet to make his greatest discovery. For his next trick, Hubble would show that the Universe had not existed forever, as most astronomers believed, but that it had a beginning.

The man who laid the groundwork for Hubble's greatest discovery was Vesto Melvin Slipher, an astronomer at the Lowell Observatory in Flagstaff, Arizona. Ever since 1912, well before anyone knew about galaxies, Slipher had been painstakingly measuring the patterns in the light from spiral nebulae.

Just as in sunlight, the light from these nebulae was a mixture of colors. Each color corresponded to a particular wavelength of light:[3] the longest was red and the shortest blue. With the aid of a prism—a triangular wedge of glass—it was possible to spread out the colors into an ordered sequence known as a spectrum.

In the nineteenth century, astronomers had found that the rainbow-like spectra of the Sun and of the nebulae were interrupted by ugly dark lines. Colors were missing. It was soon realized that these "missing" colors had been removed, or absorbed, by gases in their atmospheres. From the pattern of dark lines it was possible to actually identify the gases that were doing the absorbing—gases like helium or nitrogen or oxygen.

Slipher's great triumph was to perfect a technique for photographing the spectra of extremely faint objects such as spiral nebulae. By 1917, he had studied fifteen of these with the telescope at Flagstaff, and what he had discovered puzzled him greatly.

---

[3]Light is a wave like a wave on water. And just like a water wave it has peaks and troughs. The wavelength of any wave, whether a light wave or a water wave, is defined as twice the distance between successive peaks and troughs.

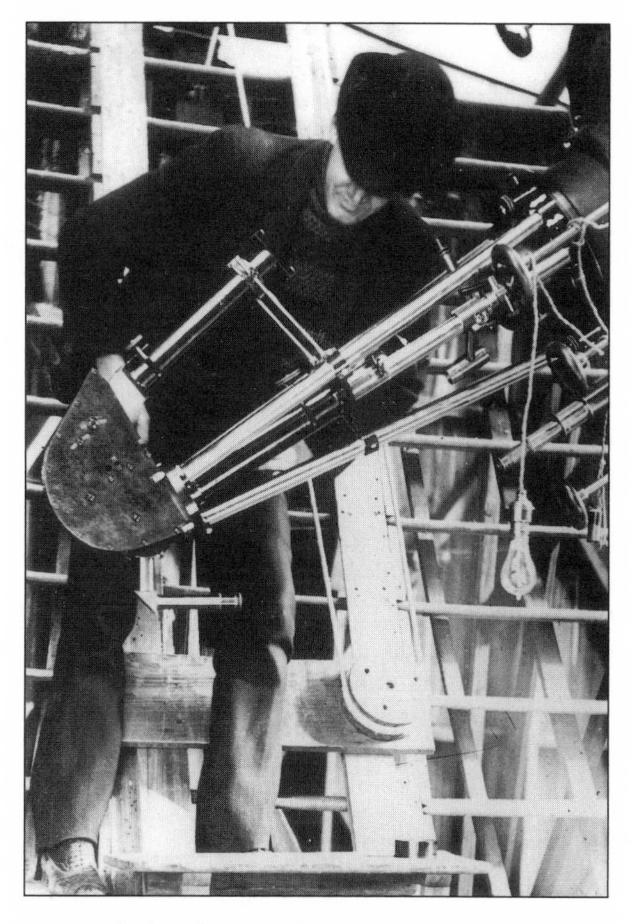

Vesto Slipher, the man who first discovered that most galaxies are fleeing from us.

In the spectra of the Sun and the stars of the Milky Way the dark lines of absorbing gases appear very close to the positions measured in laboratories on Earth when the same gases absorb light. But Slipher found that in the nebulae, the lines were shifted—usually to the longer wavelength end of the spectrum where the light was redder. In only two of his sample of fifteen nebulae were the lines shifted towards the blue end of the spectrum.

Slipher interpreted the wavelength shifts as being due to the Doppler effect, which is familiar to anyone who has noticed how the pitch of a police siren changes as it speeds across town, becoming higher as it approaches then deeper as it recedes into the distance.

As a sound wave passes, the air is alternately compressed then expanded. That is all a sound wave is: a long train of alternating "compressions" and "rarefactions" of air. The longer the wavelength—essentially the distance between one compression and the next—the deeper its pitch.

Waves from an approaching siren are "scrunched up," shortening their wavelength and making them higher pitched, while waves from a receding siren are "stretched out," deepening their pitch.

When the wavelength of light is changed, on the other hand, this causes a change in color rather than a change in pitch. So for a body coming toward us, the Doppler effect shortens the wavelength of the light, shifting its characteristic pattern of colors to the blue end of the spectrum; on the other hand, the same effect drags out and lengthens the wavelength of the light for a body moving away, causing the pattern in its spectrum to be red-shifted.

We are fortunate indeed that nature has created atoms that can make dark lines in spectra. If all the colors in a spectrum were simply shifted, we would never know. The spectrum would look the same. It would be like taking a sequence of numbers like 1,2,3, 4,5,6,7,8 . . . and shifting it one place to the left. The number 2 would replace 1, 3 would replace 2, and so on, but the sequence would still appear 1,2,3,4,5,6,7,8 . . .

But because of spectral lines, there are distinctive patterns in any spectrum. A spectrum looks like the bar code on products in the supermarket. So it is immediately obvious if the atomic bar code has been shifted.

Because thirteen of Slipher's fifteen nebulae had red shifts, this meant that thirteen were racing away from us while only two were coming our way. But this seemed to defy common sense. The nebulae were in different parts of the sky and so not connected to one another. They should therefore be moving in random directions. By the laws of chance, roughly half the nebulae should be approaching and half receding. Why should there be any pattern at all in their velocities?

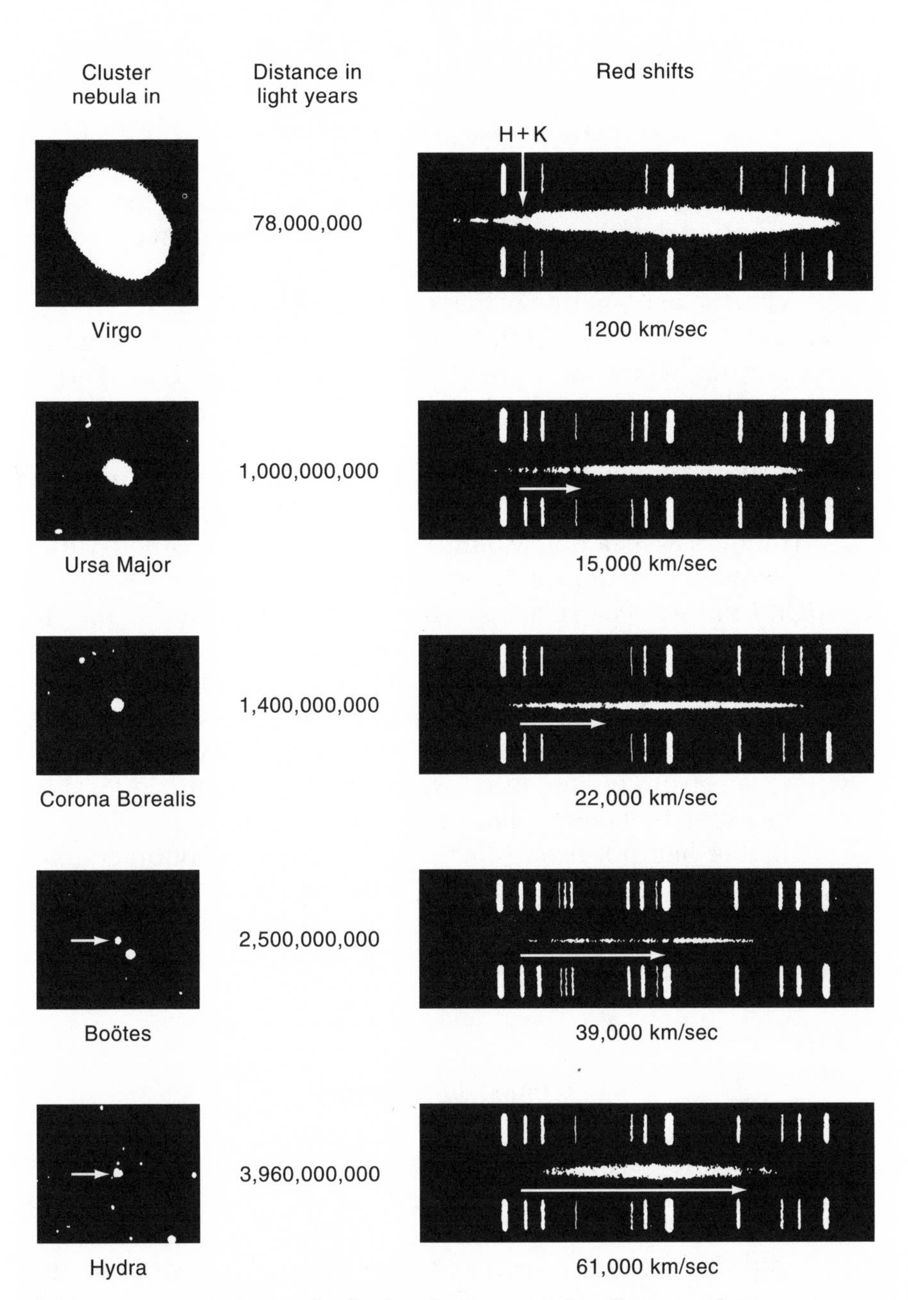

The farther away a cluster of galaxies, the faster it is hurtling away from us. Astronomers deduce the recession velocity of a cluster from the red shift of prominent features in its spectrum.

There was something else peculiar about the red shifts of the receding spiral nebulae. The shifts were very large, much larger than those of ordinary stars in the Milky Way. Taken at face value, they implied that the nebulae were receding from us at enormous speeds of thousands of miles a second.

A partial explanation of these speeds came in 1923 when Hubble discovered that the spiral nebulae were galaxies. Since they had nothing whatsoever to do with the Milky Way, there was no reason why they should be moving like stars in the Milky Way. But though the high red shifts could be swept under the carpet, there was still no explanation of why most spiral nebulae were fleeing from us.

Hubble's assistant at Mount Wilson was a man called Milton Humason, a one-time mule driver on the mountain who had taught himself to be an astronomer. On Hubble's suggestion, Humason began to extend Slipher's pioneering work. He measured the velocities of the faintest, and therefore most distant, galaxies that could be seen with the 100-inch telescope, and very soon confirmed that Slipher was absolutely right. Every single galaxy whose spectrum he measured was receding from us, some at incredible speeds of tens of thousands of miles a second.

Hubble had not been idle while his assistant photographed spectra. He had been painstakingly measuring the distances to Humason's galaxies, assuming that they were all of the same brightness so that the fainter ones really were further away than the brighter ones.

A beginning to time

In 1929, while staring at the data, it dawned on Hubble that the red shifts of the galaxies were not random at all. There was a pattern: the further away a galaxy, the faster it seemed to be hurtling into the void. In fact, the velocities of the galaxies increased in step with their distances. A galaxy that was twice as far away as another turned out to be receding from us at twice the velocity; a galaxy three times as far away was receding at three times the velocity.

This pattern would come to be known as Hubble's law.

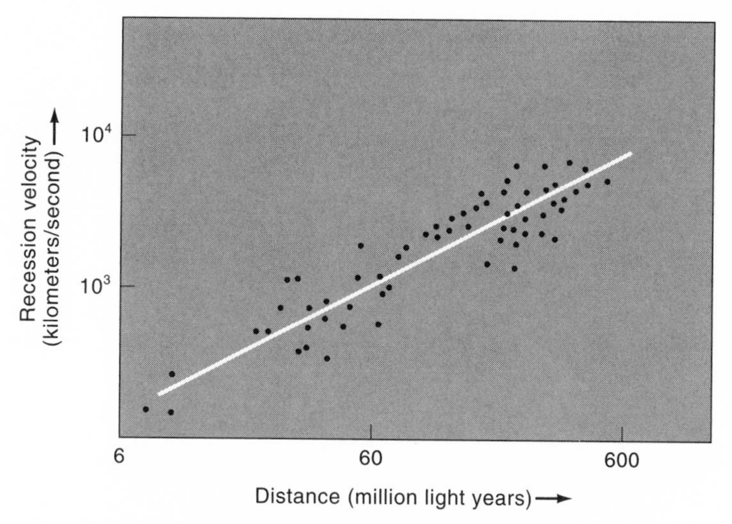

The recession velocities of spiral galaxies increase in step with their distance—a striking demonstration of Hubble's law.

The simplest and most naive explanation of what Hubble had found is that at some time in the remote past a violent explosion took place, centered on the Earth. The galaxies were blasted outward so that today when we observe them we quite naturally see them all racing away from the origin of the explosion. Those galaxies that came out of the explosion moving relatively slowly have covered the least distance while those that started off fastest have receded furthest from us.

Hubble had made the outstanding astronomical discovery of the century. The entire Universe was expanding, its constituent galaxies flying apart like cosmic shrapnel. But if the Universe was expanding, then one conclusion was inescapable: it must have been smaller in the past. There must have been a moment when the headlong expansion started: the moment of the Universe's birth.

This was the real significance of Hubble's discovery. By hypothesizing an expanding universe, he had also found that there must be a beginning to time—that although the Universe was old, it had not existed forever. By imagining the expansion running backward, like a movie in reverse, astronomers now deduce that the Universe came into existence in the Big Bang about 15 thousand million years ago. For the first time, scientists would be able to ask where the Universe—with its galaxies, stars and living organisms—had come from and where it was going. Cosmology—the most audacious of sciences—was born.

2

THE RESTLESS UNIVERSE

How Einstein missed the message in his own equations

Edwin Hubble's discovery that the Universe we live in is expanding in the aftermath of a gigantic explosion should have surprised no one. Not only had several scientists predicted it more than a decade earlier, but their predictions had also been published in the scientific literature for everyone to read. No one took a blind bit of notice—least of all Hubble.

The man who had made it possible to think seriously about the kind of Universe we live in was Albert Einstein. In 1915, he had published his theory of gravity,[1] which described the way in which every chunk of matter pulls on every other chunk. Never one to shy away from the really big problems in science, Einstein two years later applied his theory of gravity to the biggest collection of matter he could think of—the entire Universe. In doing so, he created cosmology, the science that concerns itself with the nature of the Universe we live in—where it has come from and where it is going.

According to Einstein's theory, matter does not influence other matter directly but only through the intermediary of space. This is the crucial difference between Einstein's view of the Universe and the view of his famous predecessor, Isaac Newton. To Newton, space was simply the backdrop against which the cosmic drama was played, but in Einstein's theory, it has a far more active role.

Einstein's essential idea is that space is malleable—it can be warped or curved by the presence of matter. Now warped space is

[1] Also known as the general theory of relativity.

Edwin Hubble (right) and Albert Einstein (second from left) on Mount Wilson in 1931.

a hard thing to imagine, but though we cannot visualize it, we can gain some insight into its most important properties by thinking of it as a pliable rubber sheet. If a heavy ball bearing is placed on such a sheet, it creates a depression or valley around it.

In the same way, a massive body like the earth creates a valley in the space around it.

Now imagine placing a second ball bearing on the rubber sheet. Since the first ball bearing rests at the bottom of the valley it has created in the rubber sheet, the second ball bearing will naturally roll down towards it.

In the same way, small bodies in space fall into the "warped space" around the Earth.

We say that the Earth attracts other bodies with its gravitational force. But in reality, the Earth warps space and it is this warped space that affects other bodies. This is what gravity is: warped or curved space.

The whole idea can be neatly summarized in the sentence: "Matter tells space how to warp, and warped space tells matter how to move." It's all rather chicken-and-egg like, but many experiments since Einstein proposed his theory of gravity in 1915 have confirmed this is the way things work.

Einstein's blind spot

In 1917, when Einstein applied his theory of gravity to the Universe as a whole, by rights he should there and then have discovered that it was expanding. It was crying out at him from his equations. But the greatest physicist of the twentieth century did not see it. Or rather he did see it but he ignored it.

What obscured the truth for Einstein was simple prejudice. He had already decided how the Universe should be, so he was primed to ignore all competing possibilities.

Einstein had a deep-seated belief that the Universe we lived in was "static," that all the bodies in it were essentially hanging motionless in a space. After Hubble's discovery of galaxies in 1923, Einstein was able to be more specific and say that the bodies hanging motionless in space were the galaxies. It was possible for individual galaxies to wander about a little within the Universe, but not so that it changed the overall density of matter. That had to stay the same forever.

A static universe appealed to Einstein because it made things simple. A static universe could never surprise you. It would remain exactly the same throughout time. There was no need to worry about answering sticky questions about where the Universe had come from or where it was going. There was no beginning; there was no end. The reason the Universe was the way it was, was because that was the way it always had been.

But when Einstein applied his theory of gravity to the Universe, he found that the galaxies seemed to have a restless need to be on the move. The reason is clear. Every galaxy is pulling on every other galaxy with the force of gravity, so the net effect should be to pull all the galaxies together.

This was all a worry to Einstein, but his belief that the Universe must be static was so great that he was not going to let go easily. It was very difficult indeed to make the Universe static. To salvage the idea, Einstein had to resort to mutilating his elegant equations. He invented a mysterious force of cosmic repulsion. The force could be felt only over enormous distances, which is why we had not noticed it before. It counteracted the gravitational force which was remorselessly pulling all the galaxies together.

There was no evidence whatsoever that such a peculiar force existed, but if it did, Einstein reasoned, it would stop all of creation from collapsing in on itself. The static universe would be rescued from a premature grave.

If this sounds contrived, that's because it was. In fact, there were much more natural solutions to Einstein's equations, though ironically, it was left to others to see the truth in his original equations.

The evolving Universe

One of the first people to accept Einstein's theory of gravity was a friend of his, the Dutch astronomer Willem de Sitter. In 1917, he, too, had applied the theory to the entire Universe. But unlike Einstein, he did not insist that the density of the Universe remain constant for all time. Instead, he looked at the equations with a slightly more open mind.

De Sitter discovered an entirely different design for the Universe, which also obeyed Einstein's equations. In one way it was greatly at odds with the Universe we live in because it was completely void of matter. But it had another property that was remarkably like the Universe we live in: its space was expanding.[2]

If two particles were placed somewhere in this empty universe, they would move steadily apart as the space between them

---

[2]The irony was that de Sitter had been looking for a static universe that obeyed Einstein's equations and that was less contrived than Einstein's.

Aleksandr Friedmann, the man who made the Universe expand, in 1922.

expanded. If a large number of particles were scattered throughout such a universe, the general expansion of space would steadily increase the distance from one of them to any other. In fact, every particle would recede from every other particle at a speed proportional to the distance between them. In de Sitter's universe, a law like Hubble's would apply quite naturally.

The red shifts in the light of distant galaxies have a rather simple explanation in such an expanding universe. Rather than being Doppler shifts, they arise because in the time that light from a distant galaxy has been travelling across space to us, the Universe has inflated in size, stretching the wavelength of light along with it. Imagine drawing a wiggly wave on the surface of a balloon, then inflating it. This illustrates how light is stretched in wavelength, or red-shifted.

Apart from having a rather interesting expansion law, de Sitter's universe did not have much going for it. After all, it was empty of matter. But in 1922, a Russian astronomer called Aleksandr Friedmann at the University of Petrograd rectified this. He discovered a whole class of universes that obeyed Einstein's equations and which, like the real Universe, contained particles of matter.

Friedmann found that his universes would almost certainly not be motionless—they would change their appearance with time, by either expanding or contracting. In the expanding universes, the particles of matter naturally obeyed Hubble's law.

Astronomers call such universes, that change with time, "evolving," to distinguish them from static universes that stay the same. The evolving universes of Friedmann were discovered independently five years later by Georges Henri Lemaître, a Catholic priest turned astronomer.

A characteristic feature of the universes of Friedmann and Lemaître was that they began with a violent expansion from a small and highly compressed state: a Big Bang. Particles of matter were born on the move and have been flying apart ever since.

Lemaître went on to speculate about what had actually caused the explosion at the beginning of the Universe. He knew about the phenomenon of radioactivity in which an atomic nucleus disintegrates releasing a lot of energy. It was therefore natural for him to suppose that the Universe had begun when a giant "primeval atom" exploded sending all of creation flying apart. There was little evidence for this but at the time no one had a better idea.

Einstein's biggest blunder

When Hubble discovered that the Universe was expanding he vindicated what Friedmann and Lemaître had been saying for years. Our Universe is evolving. It began in a Big Bang and has been expanding ever since. Questions like what the Big Bang was and what happened before it might be difficult but they would simply have to be faced. The plus was that a universe that was forever changing was bound to be richer in possibilities than a static cosmos, frozen in eternal immobility.

When Einstein learned of Hubble's discovery he realized his error in inventing his cosmological repulsion. Immediately, he renounced it, calling it "the biggest blunder of my life."

In actual fact, Einstein's static universe could never have worked and this was shown by the British astronomer Arthur Eddington in 1930. A static universe was inherently unstable, balancing precariously on the knife edge between expansion and contraction. The slightest nudge would have sent it careering either way.

But in Einstein's defense, it should be said that when he applied his theory of gravity to the Universe in 1917, nobody even knew that the major constituents of the Universe were galaxies. He can be forgiven his uncharacteristic lapse.

Did the Big Bang happen in our backyard?

Naively we thought of the Big Bang as a titanic explosion centered on the Earth, in which the galaxies were blasted apart like cosmic shrapnel. But the equations of Friedmann and Lemaître describe something quite different. If the Big Bang was an explosion it was an explosion unlike any other.

For one thing, when a bomb goes off, shrapnel is blown outward into a void that already exists—the surrounding air. But no such void existed before the Big Bang. There was literally nothing. The Big Bang created everything, and that included empty space, matter, energy and even time. As soon as it was created, the Universe began expanding.

If you have trouble visualizing this, don't worry. The Big Bang was unique: a one-off event. There is nothing in our everyday experience to compare it to; words are inadequate.

Another major difference between a familiar explosion and the Big Bang is that the Big Bang happened everywhere at once. It would have been impossible to point to a place and say that it was the center of the explosion, in the way that you can point to the place where a bomb went off. Fifteen billion years ago, every particle of matter was simply set in motion rushing away from every other particle of matter.

An explosion that occurs everywhere in space has an important consequence. It gives every observer in the Universe the illusion that they are at the center of the explosion. So although we see every other galaxy rushing away from us, it does not mean that we are in a privileged position at the center of the Universe.

The best way to see why this is so is to imagine the Universe as a rising cake with raisins representing the galaxies. There are flaws in this picture—for instance, a cake has an edge whereas the Universe goes on forever—but, by and large, the picture works.

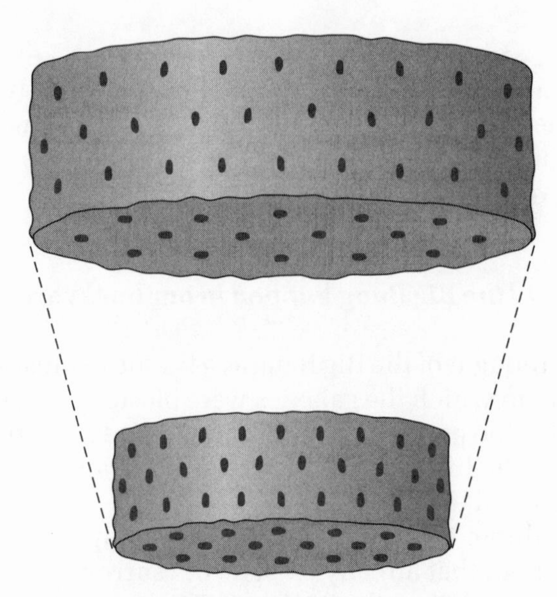

Imagine the Universe as a rising cake with raisins representing galaxies. As the cake expands, the raisins move away from each other. In fact, every raisin sees every other raisin receding. So it is with galaxies in the expanding Universe. Every galaxy sees every other galaxy rushing away.

As the cake rises, the cake mixture expands in all directions, driving the raisins farther and farther apart. Now if you were to look at the view from any raisin—it doesn't matter which—you would always see every other raisin moving away. In the same way, it would not matter if we lived in the Andromeda galaxy or a galaxy at the limit probed by our most powerful telescopes: the galaxies would always appear to be rushing away from us just as they do from the Milky Way. In our expanding Universe everyone sees the same view, and everyone thinks they are at the center of creation.

Astronomers have a name for this feature of the Universe, namely that no place is more special than any other. They call it the Cosmological Principle. It is a natural extension of a principle formulated by the great sixteenth-century Polish astronomer, Nicolaus Copernicus. He lived at a time when Ancient Greek ideas of an Earth-centered cosmos still flourished. But his observations showed that the Earth revolved around the Sun, not the other way around. The Copernican Principle can be simply stated: our place in the Universe is in no way special. The Cosmological Principle is a natural extension of this idea from the sixteenth-century Universe consisting of the Sun and planets to the twentieth-century cosmos crowded with galaxies.

Why Hubble's law must be true

It turns out that Hubble's law is a natural consequence of living in an expanding universe where the Cosmological Principle applies. The speed of a receding galaxy has no option but to be proportional to its distance. To see why, think of three galaxies A, B, and C, which happen to lie in a straight line. Let's say the distance between A and B is the same as the distance between B and C.

Now imagine that B is receding from A at 100 miles a second. This means that C must be receding from B at 100 miles per second as well, because we know the Universe looks the same from every point. This is the Cosmological Principle.

How fast is C receding from A? Well, it must be 100 miles per second plus 100 miles a second—200 miles a second. So C, which is twice as far away from A as B, is receding at twice the speed.

If we extended this reasoning to all galaxies in the Universe, we would find that a galaxy three times as far away as another would be moving three times as fast, and so on. This is precisely the expansion law Hubble discovered in 1929. It turns out that if the Universe is expanding and it also looks the same from every point, this expansion law has to be true.

Why is the sky dark at night?

Although Einstein wanted the Universe to be static and infinite; the evidence that this is not so has always been around for people to see. In fact, evidence can be found in the simple observation that the sky is dark at night.

If the Universe stretched forever in all directions with stars marching on, rank after rank, out to infinity, then in every direction you looked out from Earth you would see a star. Between the bright stars in the sky there would be fainter stars, and between them fainter stars still, on and on forever, so that there would be no gaps at all between the stars. Since every line of sight from the Earth would sooner or later strike the surface of a star, the entire night sky would appear as bright as the surface of a typical star, a result in spectacular disagreement with what we actually observe.

It was the German astronomer Johannes Kepler—renowned for discovering the laws that govern the motion of the planets around the Sun—who first pointed out this apparent paradox in 1610. Other astronomers, including Edmund Halley, the man the famous comet was named after, also recognized the contradiction between theory and observation, but it was the German astronomer Heinrich Olbers who popularized it in the early nineteenth century. Today, it is generally known as Olbers's paradox.

Another way to see the argument is to think of the Universe as made of concentric shells of space rather like the skins of an onion. Sheer distance will make the stars in a remote shell appear much fainter than the stars in a shell close to the Earth. But though these distant stars may be individually fainter, there will be more of them, because the distant shell will be larger. In fact, it turns out that no matter how far away a shell is, the number of stars will always compensate for their faintness, so that each successive shell will contribute the same amount of light. Since there are an infinite number of such shells in a never-ending Universe, the brightness of the sky should therefore be infinite!

Actually, this is not quite right. The stars may seem no more than pinpricks but in fact they are tiny discs (though no telescope is powerful enough to discern them). Because of this, nearby stars will block out the light from more distant ones behind them. When this effect is taken into account, a slightly less ridiculous answer is obtained: the night sky should not be infinitely bright but as bright as the average star.

Most of the stars in the Universe—about seventy per cent—are a type known as red dwarfs, quite a bit cooler than our own Sun. So the night sky should appear completely red, as if we lived on the surface of a red dwarf. In fact, the night sky is a thousand million

million million times fainter at visible wavelengths than the surface brightness of such a star.

The fact that the sky is dark at night, an apparently trivial observation, is therefore telling us that the Universe cannot be static and filled with stars marching on and on forever.

In a Universe like ours that has undergone a Big Bang, two things stop the night sky from being bright. The first is the expansion of the Universe. Because the Universe is expanding the light coming from ever more distant galaxies is progressively more redshifted. Since red light carries less energy than blue light, the effect of this is to reduce the energy of light from distant galaxies. As a result, galaxies at great distances contribute less to the brightness of the night sky than they would if the Universe were static.

But there is another, much more important, effect in a Big Bang universe that helps to keep the night sky dark. It is the fact that the Universe had a beginning and so has not existed forever. This means that not every line of sight ends in a star, as Kepler and Olbers and the rest assumed.

To understand why this is so, you have to realize that we only see a distant star or galaxy if there has been enough time since the Big Bang for the light to have reached us. If there has not, we simply do not see it.

It all comes down to the speed of light, which, though exceedingly fast by everyday standards, is not infinite. Light travels at about 186,000 miles a second, or about 670 million miles an hour. Snap your fingers. In the time it took to do that, a ray of light could have made the round-trip between Europe and America about thirty times.

But though light is swift, the Universe is a very big place. Light takes about eight minutes to reach the Earth from the Sun, more than four years to come from Alpha Centauri the nearest star, but billions of years to reach us from the most distant galaxies. If the Sun were to wink out at this moment, we would not know about it for eight minutes. Almost certainly, the most distant galaxies have changed in the time light has taken to reach us (they may be long dead, for all we know). The finite speed of light means that as we look farther and farther into space, we see objects as they were further and further in the past.

But the finite speed of light has another consequence in a Universe with a beginning. Although stars may march on for a very long way indeed, there is a limit or "horizon" beyond which we cannot see them. There has simply not been enough time since the beginning of the Universe for their light to reach us. An analogous horizon exists around a ship at sea. It is not the end of the Universe as far as the ship's captain is concerned: it is simply as far as he or she can see.

This effect of only seeing stars or galaxies within a certain horizon is the most important reason why the sky is dark at night. Today, the distance an arbitrary line of sight must hypothetically extend before intercepting the surface of a star greatly exceeds the distance to the horizon.

So in a Big Bang universe the sky is dark at night because of the Universe's finite age and, to a lesser extent, its expansion. In fact, the neat thing about a Big Bang universe is that these two effects go hand in hand. The expansion was caused by the Universe exploding into being relatively recently in a Big Bang.

There is a footnote to all this. It turns out that although Kepler, Olbers and the rest were right to point out that it was a great mystery why the night sky was dark—the mystery explained in a Big Bang universe—they were wrong to go on and say that the night sky should be as bright as the average star. What they had forgotten was that stars do not live forever. They run out of fuel and wink out, usually within 10 billion years or so. But as the astrophysicist Ed Harrison pointed out in 1964, it would take the stars in the Universe something like 100,000 million million millon years to fill space with enough radiation to make the night sky appear as bright as the surface of the average star. So Olbers's paradox never really was a paradox: it was a red herring.

Nevertheless, it generated a lot of hard thinking about an infinite static universe—thinking that eventually showed that such a universe could never exist. Einstein had missed the message in his own equations of gravity: the Universe would not have been possible had there not been an origin of time, followed by an expansion.

Fred Hoyle, the English astronomer who invented the term Big Bang, although ironically he never believed in the idea.

The Big Bang versus the Steady State

But the idea of a universe that always stays the same was not dead yet. So great was its aesthetic appeal that Einstein had resorted to inventing a cosmological repulsive force to keep the Universe unchanging in space and time. Others were prepared to bend the laws of physics in other ways to keep the static universe alive.

In 1948, the British cosmologists Fred Hoyle, Hermann Bondi and Thomas Gold proposed the Steady State theory of the Universe. It was based on something known as the Perfect Cosmological Principle. The Perfect Cosmological Principle went one step further than the standard version. It maintained that the Universe looks the same wherever you are *for all time*.

Hoyle and his colleagues maintained that the Universe expands at a constant rate, and that matter is created continuously throughout it to fill the voids left behind. This matter, popping into existence out of empty space, is just enough to compensate for the expansion and to keep the density of the Universe constant.

Where this matter would come from, neither Hoyle, Bondi nor Gold could say. But then nobody could say where the matter in the Big Bang came from either. For the next decade and a half, it was a two-horse race between the Big Bang[3] theory and the Steady State theory. But by the early 1960s, the Big Bang was nosing ahead.

At the University of Cambridge in England, the astronomer Martin Ryle had been carrying out a survey of radio galaxies, objects that generated intense radio waves, radiation physically identical to light waves but with a wavelength a million times longer. He was finding that radio galaxies were more common far away than they were in the neighborhood of the Milky Way. Since the radio waves from the distant galaxies had taken billions of years to reach the Earth, Ryle concluded that these objects were far more common in the remote past than they are today. In other words, the Universe had changed with time, in clear conflict with the Steady State theory.

---

[3]Ironically, it was Hoyle who coined the name Big Bang for the exploding rival of his Steady State theory. He used the term in the last of a series of five radio talks, which he delivered for the BBC in 1950.

3

THE PRIMEVAL FIREBALL

Cooking up the elements in a hot Big Bang

In 1934, when a 30-year-old Russian physicist named George Gamow stepped off a ship in New York and surveyed his new home, Hubble's discovery of the expanding Universe was already five years old. But it had not yet sunk deeply into the scientific consciousness. Although many scientists accepted Hubble's proof that the Universe had indeed begun in a titanic explosion billions of years ago, nobody seriously thought that science could say anything about what happened in the Big Bang. The idea was simply too preposterous to consider.

Such a failure of nerve is common. Scientists may scrawl arcane formulae across blackboards with reckless abandon, but deep down they find it extremely hard to believe that nature really dances to the tune of their flimsy equations. When those equations describe the birth and evolution of the entire Universe—as Einstein's did—it takes a brave scientist indeed to follow through their implications.

George Gamow would prove to be such a brave scientist.

Before emigrating to the USA, Gamow had studied cosmology under Aleksandr Friedmann at the University of Petrograd. In Cambridge he had worked with Ernest Rutherford, who had created the science of nuclear physics, and in Copenhagen with Niels Bohr, who had created our modern picture of the atom. Gamow's interests ranged far and wide, from the theory of stars to biology[1] to popular science writing. But it would be in cosmology that he would truly make his mark.

[1]Gamow was the one who guessed that short sequences of nucleic acid "bases" along DNA might form a "code" that carried the "blueprint" for the proteins of our bodies.

George Gamow, the man who guessed the Big Bang had been a blisteringly hot fireball.

Gamow was wrong about almost everything. But his achievement was immense, for he was the first person to take the Big Bang really seriously and use nuclear physics to predict what the earliest moments of creation were like. Decades later, others, following his lead, would go on to speculate about the first split second of the Big Bang.

On the face of it, what got Gamow thinking about the Big Bang seemed to have nothing whatsoever to do with the explosion at the beginning of time. In the 1930s, Gamow set out to explain where the chemical elements had come from. Where, he wondered, did oxygen and carbon come from, and iron and gold? Atoms of these elements made up everything in the Universe—our bodies, the Earth, the stars—so where did they come from?

When Gamow began thinking about this problem, astronomers already possessed an important clue. Over the years, the spectra of thousands of stars had been carefully examined. From the patterns of colors missing in each spectrum, astronomers were able to deduce which elements were absorbing the light. This enabled them to measure how common each element was in different parts of the Universe.

What they discovered was that the elements existed in roughly the same proportions absolutely everywhere. It was a clear indication that some common process had made all the elements in the Universe. Gamow guessed that originally the Universe had contained only a very simple ingredient, and that somehow all elements had been made from this ingredient. Gamow was not the first person to have this idea but he would take it further than anyone else.

The world of the atom

When Gamow started his quest, physicists knew that all elements—from hydrogen, the very lightest, right up to uranium, the heaviest—were built from just three basic building blocks: tiny particles called protons, neutrons and electrons. Every atom consisted of a "nucleus"—a clump of protons and neutrons—sitting like a sun at the center of a cloud of furiously orbiting electrons.

The key thing that made a hydrogen atom different from, say, an atom of carbon or uranium was the number of protons in its nucleus, which was precisely matched by the number of electrons in orbit. Hydrogen had just one proton and one electron, whereas carbon had six protons and six electrons. The nucleus of uranium, on the other hand, was a monster. It contained 92 protons and sat in a haze of 92 whirling electrons.

Protons and electrons were bound together by the "electrical" force between them. Electrons have a negative electrical charge and protons have a positive charge. Nobody really knows what electrical charge is, only that the electrical force between particles with different charges—electrons and protons—makes them attract each other, whereas the force between two similarly charged particles causes them to repel each other.

Neutrons have no electrical charge, which means they are unaffected by the electrical force. They keep protons separated so that they can live together inside a nucleus. Hydrogen needs no neutrons because it has a lone proton, but to keep the peace in a uranium nucleus 150 neutrons are needed. Without these, the electrical force between all the positively charged protons would simply blast the nucleus apart.

Clearly there must be another force to counterbalance the electrical force, or atoms could never exist. There is. The "strong nuclear" force provides the glue which binds together neutrons and protons inside a nucleus.

Unlike the electrical force, the nuclear force has an extremely short range. This means that protons and neutrons have to get very close together before they feel it. Once they do get close enough, it grabs them enormously tightly.

But although the nuclear force is strong, it is not overwhelming. The basic building blocks within a nucleus are able to rearrange themselves. Early in the century it was discovered that this happens naturally in some "radioactive" atoms. Their nuclei are unstable and sometimes spit out some of their neutrons and protons quite spontaneously, changing into other atoms in the process. And what nature can do, physicists soon learned to do also. In 1919, Ernest Rutherford became the first person to artificially change one element into another when he bombarded nitrogen with alpha particles—the nuclei of helium atoms—and made oxygen.

The cosmic cake mix

The idea that atoms could be changed by adding or subtracting the basic building blocks—protons and neutrons—was the clue Gamow leaped on in his search for the origin of the chemical elements.

Gamow guessed that the Universe had started off with a mix of protons, neutrons and electrons, and that all the elements had been assembled from these. One of his later collaborators would call the mix the "ylem." If the ylem were dense enough and hot enough, the protons and neutrons would start colliding and sticking together to make light elements, and the light elements in turn would collide with each other to make heavier elements.

There are many different reaction schemes you could imagine for building up the elements. But these would be determined by the known laws of physics once you fixed the initial mix. All you had to do was run through the calculations and see if at the end of the day you came out with the mix of elements we see around us today.

It was like trying to make a fruit cake without a recipe. One way would be to put together some likely ingredients and bake them in the oven. You could then compare the final cake with one bought in a shop. If the two were not quite the same, the ingredients could be modified and the cake baked again. The end result, after much trial and error, might be a perfect fruit cake.

Starting with protons, neutrons and electrons, Gamow was trying to cook up the precise mix of elements we now find in the Universe. The ylem would have to be extremely hot, that much was clear. Nuclei would stick together only if they collided at great speed, which meant at high temperature.[2] At low speed, the electrical repulsion between the protons in the respective nuclei would force them apart long before they could get close enough to be gripped by each other's nuclear force. At high temperature—and Gamow realized it needed to be a billion degrees or so—two nuclei would slam into each other so violently that they would overcome their mutual electrical repulsion and get close enough together to be held in a nuclear embrace.

But where in the Universe could temperatures of billions of degrees be found? It seemed a tall order to find a natural furnace that could reach a temperature of billions of degrees, and so forge all the chemical elements.

The magic furnace

Gamow's great insight was to realize that the entire Universe must have been such a furnace when it was very young.

If it were somehow possible to run the expansion of the Universe backwards, like a movie in reverse, we would see it get hotter

[2]That is all temperature is: a measure of how fast the microscopic particles that make up a body are moving.

as it got denser, just as the air in a bicycle pump heats up when it is compressed. Gamow was the first to realize that the Big Bang must have been a "hot" Big Bang.

Gamow envisaged the early Universe as a seething mass of protons, neutrons and electrons compressed into a tiny, tiny volume. Something then triggered this mass suddenly to start expanding and cooling, and as it did so "nuclear reactions" among the basic ingredients formed all the elements. All this would happen in the first few minutes after the Big Bang before the expanding fireball cooled so much that nuclear reactions could no longer go on.

Gamow tried to bake the elements from several different mixes. One of his ideas, for instance, was that the ylem was a "superdense" object made of protons and neutrons. This then broke up, like the primeval atom Georges Lemaître had proposed in 1931, and the huge amount of energy released had heated the mix to billions of degrees.

To Gamow, the Big Bang was what happened when the ylem disintegrated spontaneously. Now, where the ylem had come from, and what had triggered it to break up, Gamow had no idea. Like all scientists, he was trying to answer one question at a time.

The primeval fireball

Gamow realized early on that the ylem would not contain particles alone. Matter at any temperature gives out radiation, and the hotter it gets the more energetic the radiation. At a temperature of a billion degrees, matter produces intense gamma rays—enormously energetic radiation with a wavelength far shorter than visible light.

The early Universe therefore would be a brilliantly bright fireball.

In such a fireball, light radiation could not travel any appreciable distance the way it can in today's Universe. The fireball would contain large numbers of free electrons and these would greatly hinder its progress. Free electrons are particularly good at absorbing radiation and redirecting, or "scattering," it.

Now light has the peculiar property that although it behaves like a wave when travelling through space, it acts like a stream of bullet-like particles when it interacts with matter.[3] So in the fireball each particle of light—known as a photon[4]—would repeatedly bounce off electrons.

When a photon scatters off an electron, and the photon has more energy than the electron, the photon generally loses energy to the electron. Imagine a car hitting a motorbike. Since a car generally has more energy than a motorbike, the net transfer of energy is from the car to the motorbike. On the other hand, when a photon scatters off an electron and has less energy than the electron, the photon gains energy from the electron. A similar thing happens when two electrons collide with each other. The one with more energy generally loses some to the other.

This tendency for energetic particles to share their energy with less energetic particles has an important consequence. If particles have time to interact with each other many times, then eventually a state will be reached in which the particles share the available energy as fairly as possible.

This is precisely what happened in the Big Bang fireball. Although the Universe was expanding rapidly, the interactions between photons and electrons, and between electrons and electrons, were proceeding at a much faster rate, so at every instant during the expansion the energy was shared out fairly between all particles.

Any system in which the particles have reached such a steady democratic state is said to be in "thermal equilibrium." Here, the word equilibrium does not mean that the energy of each individual particle is unchanging. Just as before, all the particles continue to be involved in the same game of give and take. What stays constant is the *number* of particles in any given range of energy. As fast

[3]Why light has this wave/particle nature is one of the great mysteries of science. In reality, light is neither a particle nor a wave but something for which we have no word.

[4]The shorter the wavelength of light the higher the energy of the photons. For instance, the photons of blue light have more energy than the photons of red light.

as particles are knocked out of the energy range, other particles are knocked in. Equilibrium here is therefore a statistical thing.

Now, matter in thermal equilibrium has a special place in the hearts of physicists. The reason is that it is simple for them to understand. In order to predict the properties of the whole system, they do not have to sit down and calculate the energy of trillions and trillions of individual particles all rushing about randomly. In thermal equilibrium, the statistical properties of the particles are predictable. In particular, the energy of the particles is distributed in a simple way, which depends only on temperature[5] and which physicists can easily compute. It does not matter whether the particles are gold atoms or protons and neutrons. If they are in thermal equilibrium at a particular temperature, their energy will be distributed in precisely the same way.

In reality, it is difficult to find matter in a true state of thermal equilibrium because to reach such a state a system of particles must be left to settle down for a long time. During this time, if any energy escapes or pours in from outside, then reaching equilibrium becomes harder. This means that the system of particles must be isolated from its surroundings.

But though the true state is elusive in nature, it is often approached closely. For instance, the interior of the Sun is close to being in thermal equilibrium. Deep in its interior, photons are bouncing around—undergoing repeated scatterings by free electrons—as if confined in a giant box. Only relatively few leak out through the surface and illuminate the Earth. But the best example of a state of thermal equilibrium was the Big Bang fireball. After all, it was confined in the box of the Universe, so there was no possibility of energy leaking either in or out.

Now radiation in thermal equilibrium with matter has a very special character. Just as the particles of matter have a simple distribution with energy, the particles of radiation—the photons—have a simple distribution with wavelength. The spectrum of this "thermal radiation" is as familiar to physicists as the face of Albert

[5]Strictly speaking, a temperature can be defined for a body only when it is in a state of thermal equilibrium.

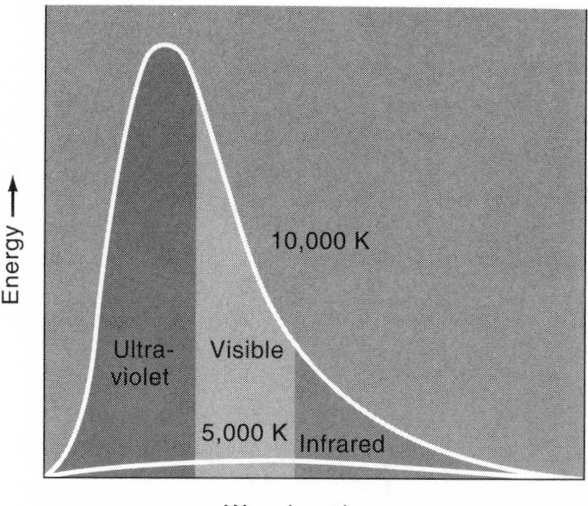

The distinctive humped spectrum of black body radiation. The energy in any range of wavelengths rises very steeply as the wavelength gets longer, reaches a peak, and then falls off steeply again. The hotter the black body, the shorter the wavelength of the peak.

Einstein. Its shape is described by a universal formula that depends only on the temperature of the material and does not depend at all on the nature of the matter that the radiation interacts with.

Radiation with an identical spectrum is emitted by a black surface that absorbs all the light that falls on it, so thermal radiation has become known as "black body radiation." This is unfortunate because the term black body, with its connotations of black holes, only confuses people. The Sun and stars are, after all, good "black" bodies. But there you are. Most physicists use the term so we are stuck with it.

A thermal, or black body, spectrum has a characteristic humped shape. The energy in any range of wavelengths rises very steeply as the wavelength gets shorter, reaches a peak, and then falls off steeply again. The hotter the black body, the shorter the wavelength of the peak. For the Sun, the peak is at the wavelength of green light.

The reason why a black body spectrum drops off at very long wavelengths is easy to see. Think of thermal radiation confined in a box with opaque walls. Wavelengths that are longer than the dimensions of the box are excluded because they simply will not fit in the box. At short wavelengths, it is necessary to appeal to the photon nature of radiation for an explanation of the drop-off in the spectrum. The shorter the wavelength of a photon, the more energy it contains. So at very short wavelengths, photons are simply too energy-hungry to be made.

The fireball radiation

In 1946, Gamow took on a research student called Ralph Alpher. In fact, Alpher was the one who first coined the word "ylem" to describe the primordial mix of neutrons, protons and electrons jumbled together in a sea of high-energy radiation from which the elements were formed.

Gamow suggested to Alpher that he calculate the quantity of various atoms that would be produced in the cooling fireball and see if they matched the quantities observed in nature. Early in the work, Gamow and Alpher were joined by Robert Herman, a graduate student from Princeton.

Alpher and Herman carried out the calculations. But they also began thinking about the fireball radiation. Like Gamow, they realized it would have the spectrum of a black body. Energy was constantly being transferred between the light and matter as electrons constantly absorbed and scattered photons.

The fireball radiation would keep its black body character even as the fireball expanded, stretching the photons to longer and longer wavelengths and cooling them. All that would happen is that the peak of the hump would shift to longer and longer wavelengths.

But Alpher and Herman realized something important that Gamow had missed. Today's Universe should be filled with remains of the heat of the fireball, greatly cooled by the expansion of the Universe.

Something enormously significant happened in the Universe about 300,000 years after the Big Bang when the temperature of the

Ralph Alpher (left) and Robert Herman putting the finishing touches on a joint paper at General Motors in 1975. The astronomical community completely ignored Alpher and Herman's 1948 prediction of a microwave afterglow of the Big Bang.

expanding fireball dropped to about 3,000°C. Until then, the Universe was a seething mass of electrons and atomic nuclei forged in the first few minutes after the Big Bang. But suddenly it was cool enough for electrons to combine with these nuclei and form atoms. Very rapidly all the electrons in the Universe would be mopped up.

The effect on the fireball radiation was dramatic. With the electrons gone there was nothing to scatter the photons of the fireball. The rapidly cooling fireball suddenly became transparent to light.

In the language of physics, photons stopped "walking" and began "flying." By walking, physicists mean that the path of the photon was reminiscent of a drunkard's progress. Each photon trav-

The "decoupling" of matter and radiation. About 300,000 years after the Big Bang, atoms formed, mopping up all the free electrons in the Universe. Since free electrons are good at "scattering," or redirecting, photons of light, the consequences were dramatic for the Universe. Photons, which until then had been forced to follow tortuous zigzag paths, were suddenly free to fly unhindered across space. We pick them up today, greatly "red-shifted" by the expansion of the Universe, as the cosmic background radiation.

elled only a short distance in a straight line before it encountered an electron and was "scattered" in another direction.

But suddenly, 300,000 years after the Big Bang, everything changed. Atoms mopped up all the free electrons so photons could fly unhindered across space.[6]

After this "epoch of last scattering," photons that had been unable to travel far in a straight line without running into an electron were suddenly able to travel unhindered. And they have been flying freely ever since, gradually losing energy as the Universe has grown in size.

At the epoch of last scattering, matter and light, which had been so intimately linked, went their separate ways. The photons of the fireball radiation have been flying across space for the past 15 billion years without ever meeting a particle of matter.

The Universe continued to expand, stretching and cooling the radiation. By now it would be only a feeble glimmer. Today, Alpher and Herman predicted, the "background temperature" would be −268°C or just 5 degrees above absolute zero.[7] The temperature of the background radiation is the temperature the Universe had long ago, but greatly reduced by the enormous expansion the Universe has undergone ever since.

---

[6]A very similar process occurs in our Sun. Photons created by nuclear reactions deep in the heart of the Sun are scattered repeatedly as they work their way up to the surface. The path they take is so contorted that they take about 30,000 years to get there! But once at the surface they are free and they take only eight minutes to fly to the Earth.

[7]Absolute zero is the lowest temperature attainable, and so has a special role in physics. When an object is cooled, its atoms move more and more sluggishly. Absolute zero (which on the Celsius scale is equal to −273.15°C) is the temperature at which they stop moving altogether.

Before the epoch of last scattering

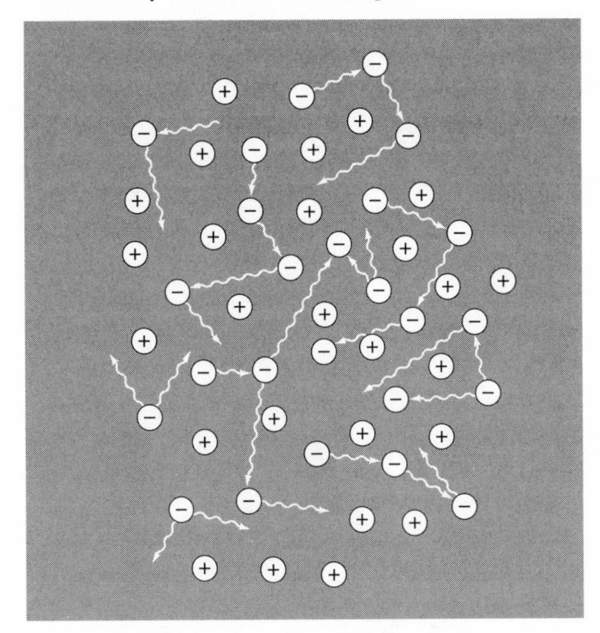

After the epoch of last scattering

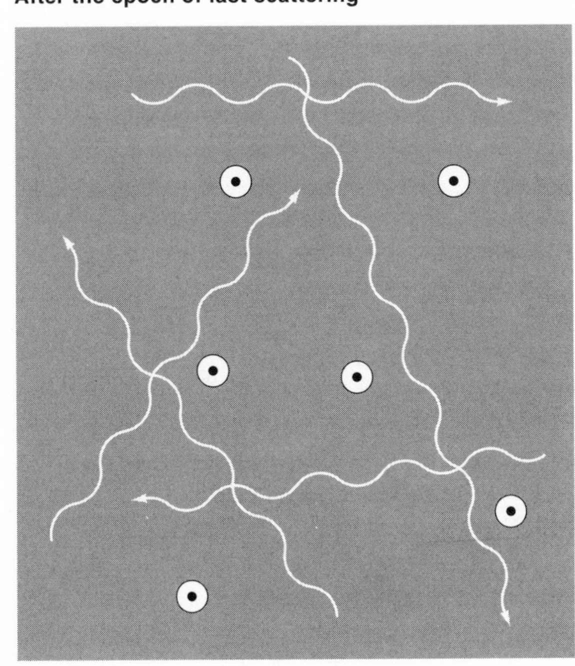

Sun

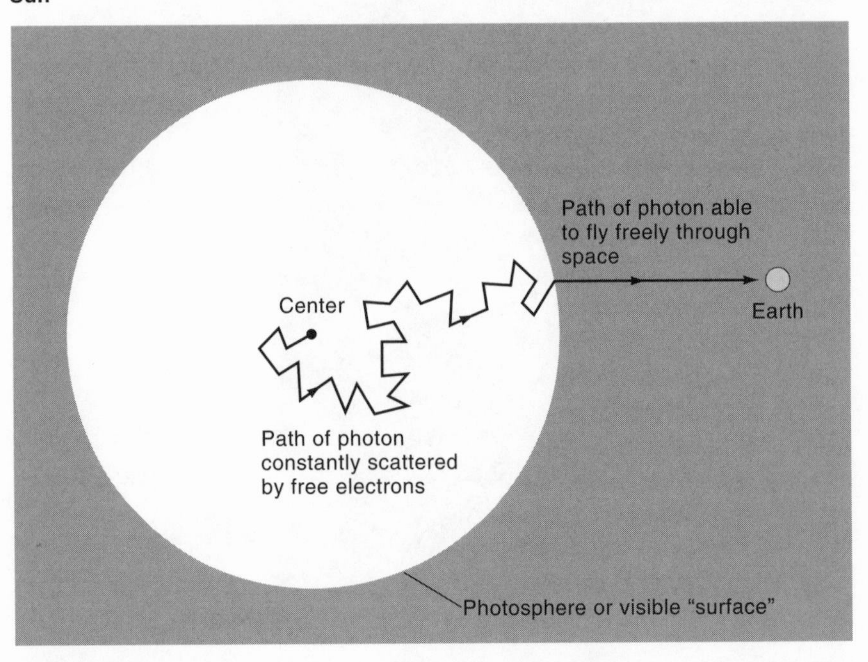

Big Bang fireball

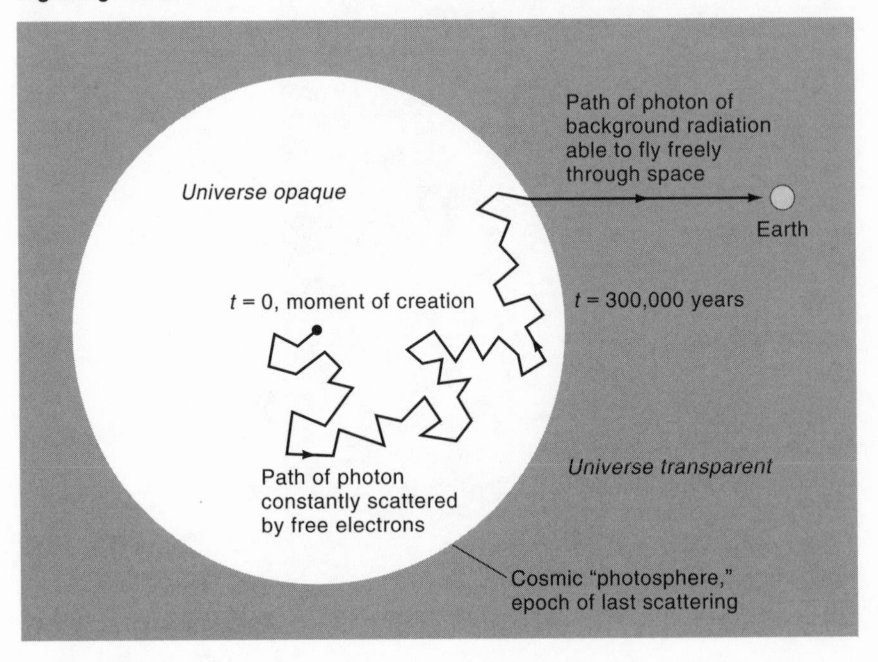

The Sun has similarities with the fireball of the Big Bang. Photons generated in the heart of the Sun are repeatedly scattered by free electrons and must zigzag their way up to the Sun's visible surface, or "photosphere." There they are suddenly free to fly unhindered across space to the Earth. In the same way, photons in the fireball of the Big Bang were repeatedly scattered and followed zigzag paths until they were suddenly free to fly unhindered across space at the epoch of last scattering—the "cosmic photosphere."

Alpher and Herman published their prediction in a paper in *Nature* in 1948. At first Gamow thought the idea unimportant. He argued with Alpher and Herman that although the Universe might be filled with this relic radiation from the Big Bang fireball, in practice it would be impossible to see it from the Earth. The problem was starlight. Gamow claimed it had the same energy density as the relic radiation, making it impossible to distinguish between the two.

Gradually, though, Gamow came round to Alpher and Herman's view. He realized that he was wrong and that the fireball radiation would have a distinct signature that would make it instantly recognizable to a telescope that was sensitive enough.

A prediction too far

Despite this, everyone proceeded to forget about Alpher and Herman's prediction. It would drop out of scientific sight for twenty years. One reason for this was that Alpher and Herman were themselves unaware that in the 1940s, there already existed telescopes capable of searching for the cooled remnant radiation from the Big Bang. In the mid-1950s, Alpher and Herman and their colleague James Follin talked to radio astronomers at the National Research Laboratory and at the National Bureau of Standards about actually looking for the relic radiation. But they were told that the available technology was simply not up to detecting such weak relic radiation.

But the most important reason for the prediction being forgotten was that Gamow's theory of how heavy elements were made was wrong. The theory worked well for helium—the simplest element after hydrogen. It predicted that about 25 percent of the material emerging from the Big Bang should be helium. This is in extremely good agreement with what astronomers find when they study stars.

But the theory failed miserably when it came to producing any heavier elements. The early Universe simply did not stay hot and dense long enough for successive thermonuclear reactions to build up elements such as carbon and iron.

As Fred Hoyle and his colleagues were to prove in the late 1950s, virtually every element heavier than helium has been manufactured since the time of the Big Bang—by reactions in the hot interiors of stars.

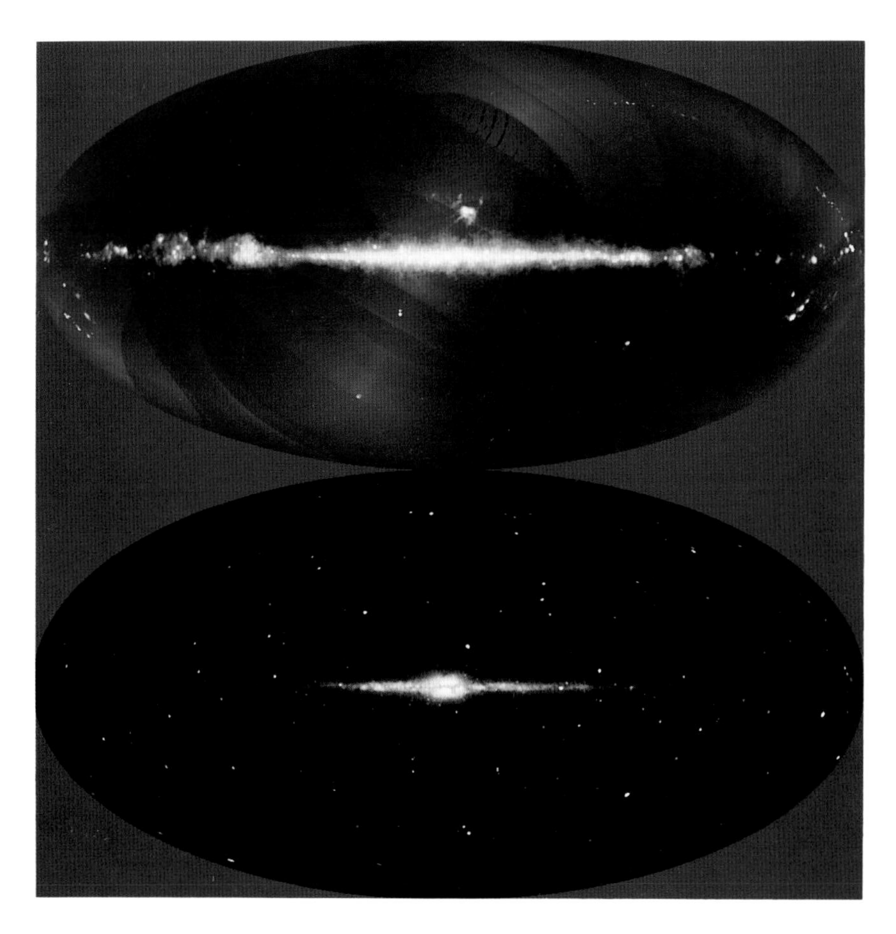

Color Plate 1 (*see Chapter 1, page 13*)
The sky at far-infrared wavelengths, as seen by one of the instruments on board the Cosmic Background Explorer satellite (COBE). The dominant source of light at such wavelengths is cold dust in vast clouds of gas and dust floating between the stars. Those clouds trace out the disc of our Galaxy, the Milky Way, which appears edge-on from our vantage point within the disc. The center of our Galaxy, a distance of 28,000 light years away, is at the center of each image.

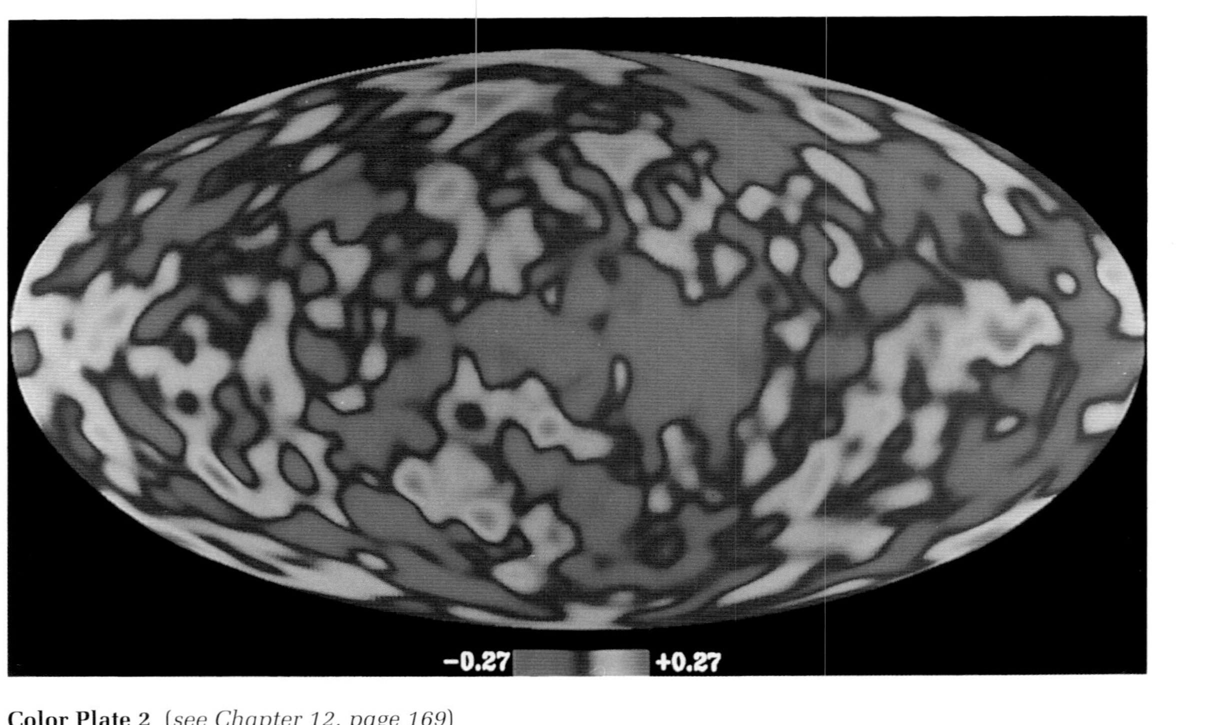

Color Plate 2 (*see Chapter 12, page 169*)
A microwave map of the whole sky made from one year of COBE data. The Milky Way is positioned horizontally across the middle of the map. Red indicates areas that are 0.01 per cent warmer and blue indicates areas that are 0.01 per cent colder than the average sky temperature of 2.73 degrees above absolute zero. Most of the red and blue patches are in fact caused by instrumental "noise." However, computer analyses indicate that faint "cosmic ripples" are also present.

4

TAKING THE TEMPERATURE OF THE UNIVERSE

The search for the fireball radiation

George Gamow seemed to have gone down a blind alley with his idea that most of the Universe's heavy atoms were made in a hot Big Bang. But in the early 1960s, a physicist at Princeton University also concluded that the early Universe had to be hot. He was unaware of Gamow's work and he had come to this conclusion for an entirely different reason. Instead of trying to build up the elements, he was trying to destroy them.

Bob Dicke was a phenomenally prolific scientist. He had trained as an atomic physicist, but he had gone on to develop an alternative to Einstein's theory of gravity and to carry out experiments to prove that Newton's law of gravity was right to an unprecedented level of accuracy. During the war, Dicke had been one of the key figures in the development of radar at the Massachusetts Institute of Technology's Radiation Laboratory.

Dicke was also interested in cosmology. But the Big Bang theory unsettled him, particularly its contention that 15 billion years ago the Universe simply popped out of nothing and started expanding. He wanted to know what happened before the Big Bang. Most scientists simply shrugged their shoulders when asked this question and said science could never answer it, but Dicke thought this a terrible cop-out. He searched for a more satisfying theory—one with fewer loose ends than the conventional Big Bang. And what he came to embrace was the idea of the oscillating, or "bouncing," universe.

Bob Dicke (right) holding the "shaggy dog"—a microwave absorber used for calibration—in front of an early "Dicke radiometer." The radiometer picked up microwaves at a wavelength of 1.5 centimeters. It was developed near the end of the Second World War at the Massachusetts Institute of Technology's Radiation Laboratory.

A giant beating heart

To Dicke, the Universe was like a giant beating heart that had been swelling and contracting throughout eternity. The reason that all the galaxies appeared to be rushing away from us, said Dicke, was simply that the human race had appeared on the cosmic stage just when the Universe was undergoing one of its swelling, or expansion, phases.

But even at this moment, the expansion was being braked by the gravitational pull of every galaxy on every other galaxy. In the future, Dicke predicted, the expansion would be slowed to a standstill, then completely reversed. All of creation would embark on a runaway collapse until matter was crushed to the maximum density possible. It was from just such a compressed state—a Big Crunch—that the Universe around us was "rebounding" today, claimed Dicke.

The great appeal of the oscillating universe was that it dispensed with the creation event and all its unsettling problems. The Big Bang was not unique. It was simply one explosion in a long line of titanic explosions stretching back through the mists of time.

The oscillating universe, like the Steady State universe, neatly sidestepped the sticky problem of what happened before the Big Bang. There had been another Big Bang. And before that another. The Universe had no beginning. It had been pulsating throughout eternity.

But there was still one loose end that Dicke needed to tie up. Since Gamow's failed attempt to make the elements in the Big Bang, Fred Hoyle and his colleagues had shown that the Universe's heavy elements had been built up from hydrogen in the furnaces at the heart of stars. Their theory was so successful in predicting which elements should be common and which should be scarce that few people doubted that it was largely correct. In fact, in the early 1950s astronomers had found that old stars did indeed contain fewer heavy elements than young stars, which was just what you would expect if as time went on stars made more and more heavy elements deep in their cores.

But if the Universe began as mostly hydrogen, and stars then cooked some of it into heavy elements, what had happened to the heavy elements that had been made during the Universe's previous cycle of expansion and collapse? There must be a process that destroyed all the Universe's heavy elements between the Big Crunch at the end of a phase of contraction and the Big Bang at the start of the next expansion.

Dicke realized that extreme heat would do the job nicely. During its compression, the Universe must have been very hot—at least a billion degrees. At such a temperature, the heavy atoms would have been slammed together so violently that they would have disintegrated into hydrogen. Every last trace of the previous era of cosmic history would be erased. The Universe would start the next cycle without any heavy elements.

An unavoidable consequence of such a hot phase in the early Universe was intense radiation. Dicke, like Gamow before him, concluded that the early Universe must have been a blindingly bright fireball.

A universal microwave background

Ironically, Dicke had wanted to break down heavy elements while Gamow had wanted to build them up. It was doubly ironic because both Gamow and Dicke were right about the existence of the fireball radiation—but for the wrong reason!

Like Gamow, Dicke wondered what would have become of the fireball radiation. He realized that the expansion of the Universe would have cooled the radiation, continually stretching out the wavelength of its photons and sapping them of energy. Instead of having a temperature of a billion degrees, the relic of the fireball should by now be only a tepid glimmer barely a few degrees above absolute zero. Instead of appearing as gamma rays, it would appear as short-wavelength radio waves.

But Dicke realized something that Gamow, Alpher and Herman had not: that there was a good chance of detecting such radiation in the Universe today.

Working in Dicke's "gravity group" at Princeton were two young physicists—David Wilkinson and Peter Roll. "One day Dicke burst into our lab," recalls Wilkinson. "He said, 'Gee, you know there might be this relic radiation in the Universe.' "

Wilkinson and Roll were intrigued by the possibility of mounting a search for the radiation from the Big Bang. The relic radiation would have two unique and striking characteristics. First, because it permeated every pore of space, it would appear to be coming from absolutely everywhere in the sky. Secondly, it would have the spectrum of a black body.

By now the radiation would be cold. It would appear brightest at short radio wavelengths between about a centimeter and a meter. These are known as microwaves. You did not need a big telescope to see the radiation since it would appear to be coming from everywhere in the sky. All you needed to take the "temperature of the Universe" was a small purpose-built radio telescope.

The first sensitive radio receivers to operate around about a centimeter in wavelength had been built for radar during the war. Radar equipment needed to be made small to fit into aircraft so there had been a major effort to make it operate at short, microwave wavelengths. Dicke was the one who in 1946 invented the instrument that became the standard for measuring microwaves from the sky.[1]

In the spring of 1964, Wilkinson and Roll started building such an instrument to look for what they had now called the "primeval fireball."

An idea that snowballed

At the same time Dicke set Wilkinson and Roll looking for the Big Bang radiation, he set a young Canadian theorist thinking about how it might be possible to estimate the present temperature of the fireball.

Jim Peebles had been working as a graduate student in particle physics ever since arriving at Princeton from the University of Manitoba. Undoubtedly, he would have stayed in that field if it had not been for a chance conversation with a fellow Manitoban, Bob Moore, who was in the year ahead of Peebles. "Bob told me that the research seminars of a faculty member he was working with—Bob Dicke—were much more interesting than what I was doing," says Peebles. "I went along to some of them—and Bob was absolutely right."

Peebles quickly learned about Dicke's idea that the early Universe had been a searing hot fireball and that the observable consequence of this might be the detection of the leftover radiation. "It was a good idea," says Peebles. "And like all good ideas it sparked a whole chain of thoughts."

Peebles immediately went to work on the implications of a hot Big Bang. The first thing he realized was that helium and a few

[1]To this day, radio astronomers refer to it as a "Dicke radiometer."

Jim Peebles, the Canadian theorist who explored the consequences of a hot Big Bang in 1965, not knowing that Ralph Alpher and Robert Herman had covered the same territory seventeen years earlier.

other elements would be produced in abundance in the Big Bang. Soon Peebles had worked out how much helium you would expect to be made and how this amount was related to the present temperature of the Universe.

What Peebles found was that about 25 per cent of the mass of the Universe should be helium. At the time he was unaware that this was precisely the helium abundance astronomers had found in many stars. "My knowledge of astronomy was exceedingly limited," he says. But earlier Peebles had written a scientific paper on the structure of Jupiter in which he had concluded that about 25 per cent of its mass had to be helium. He looked up the figure for the Sun and found that it, too, was about the same. "It was at least reassuring that I could make the numbers for the Big Bang come out consistent with what we know for the Solar System," says Peebles.

In fact, he had solved one of the great unsolved problems of astrophysics: why there was so much helium in the Universe. Although Fred Hoyle and his colleagues had proved beyond a doubt that most elements had been forged in the furnaces of stars, helium remained a big puzzle. There was simply no way that stars could have turned 25 per cent of the matter in the Universe into helium since the Big Bang. Even Hoyle was coming around to the idea that the elements must have been made in two places: the heavy elements in stars and the light elements like helium somewhere else.

Of course, Gamow had already located that place—the fireball at the start of the Universe—but because his theory was unable to produce the rest of the elements it had been discredited. It turned out that nature was not simple. The elements were not built *either* in stars *or* in the Big Bang: they were made in both places. When Gamow's work had been tossed aside, the baby had been thrown out with the bath water.

At his first ever colloquium on the subject of the hot Big Bang, Peebles told his audience that if the whole thing hung together the temperature of the relic radiation should be about ten degrees above absolute zero. "I didn't realize that Alpher and Herman had got a similar answer from a similar line of reasoning sixteen years earlier," says Peebles.

But despite the enthusiasm with which Peebles had explored Dicke's idea, he did not have high hopes that Wilkinson and Roll would actually find the Big Bang radiation. "I'm never optimistic," he says. "The hot Big Bang was simply an interesting thing to play with. I suppose I was counting on them not finding anything and considering the implications of a negative result."

The telescope in the pigeon coop

While Peebles theorized, Wilkinson and Roll got on with the job of building a telescope to look for the cooled remnant of the Big Bang fireball. They had decided to look for the radiation at a wavelength of three centimeters. Equipment was readily available because this was a common radar wavelength, known as X-band. The wavelength had the added advantage that it was one at which water vapor in the atmosphere would not be glowing too brightly. Also, the tenuous halo of gas that was known to surround the Galaxy and fill most of our sky with a background glow would not be too much of a problem either.

Wilkinson and Roll built their apparatus on the cheap, buying most of the parts they needed from army surplus stores in Phila-

David Wilkinson (with the screwdriver) and Peter Roll with their telescope in a pigeon coop on the roof of Princeton's geology building in the summer of 1965. They had hoped to be the first to measure the temperature of the Universe but had been scooped by two young radio astronomers at Bell Laboratories.

delphia, a short 45-minute drive from Princeton. They even made use of vacuum tubes, which glowed when electricity throbbed through them. "It was just at the end of an era," says Wilkinson. "Transistors hadn't quite come in."

"Neither Peter nor I knew anything about microwaves," says Wilkinson. "But Dicke knew a lot of course. We would chat with him, go off and build something in the lab, then show him what we'd done."

"Essentially, they were building the same kind of instrument I had built at MIT during the war," says Dicke. "I gave them advice and they went and did all the work of soldering."

For the site of their experiment, the two astronomers selected the roof of Guyot Hall, Princeton's geology building. "It was fine for our purpose because apart from a few towers, its roof was flat," says Wilkinson. They began assembling the antenna on a piece of plywood in a disused pigeon coop.

The heart of the apparatus was the "antenna." An antenna is simply the name given to any device that collects radio waves from the sky. For instance, a television aerial is an antenna; it collects radio waves from a TV transmitter. Other examples of antennas are the giant bowl-shaped dishes used by astronomers to pick up faint radio signals from distant galaxies.

When radio waves impinge on an antenna, they drive tiny electrical currents in its metal structure. It is by recording these currents that a radio telescope measures the strength of the radio waves.

The best type of antenna for collecting microwaves is simply a metal funnel, commonly known as a "horn." Wilkinson built his from four sheets of copper that he soldered together; it looked rather like a square trumpet, six feet long. The microwaves from the sky were collected by the flared opening, which was about a foot square. The horn then funnelled them down to a "receiver," the complicated electronic bit that actually detects radio waves. All TVs have receivers built into them. Wilkinson and Roll's was where all their glowing vacuum tubes went.

The design of the antenna was crucial to Wilkinson and Roll's experiment. All antennas are designed so that they pick up radio waves coming from only a small area of the sky while ignoring everything else. For instance, a TV antenna must pick up radio waves from the TV transmitter it is pointed at and not radio waves from other places—for instance, other TV transmitters.

But though most of the radio waves an antenna picks up come from where it is pointing, some radio waves from other directions always manage to leak in. These get into an antenna because they are able to bend round corners just like sound waves.[2] The corner in this case is the sharp metal edge at the horn's flared opening. Unwanted radio waves come from sources such as the ground, the Earth's atmosphere and the components of the radio telescope it-

[2]It is only because sound waves do bend round corners, for instance buildings, that we can hear people shouting even when they are out of sight.

self. Any material that is above absolute zero naturally produces radio waves. The common denominator is electrons. All materials—even blocks of ice—contain electrons jiggling about inside, and jiggling electrons give out radio waves. In fact, the hotter a material is the faster its electrons jiggle, and the stronger the radio waves it broadcasts.

Being able to distinguish between the signal from space and other unwanted signals is the major problem that radio astronomers face.

It is not a serious problem with a TV aerial, because the unwanted radio waves are so much weaker than those from the transmitter. But Wilkinson and Roll were wanting to measure the coldest thing in the Universe, so picking up unwanted radio waves from hotter bodies nearby was an enormous source of worry.

The Big Bang radiation would be only a few degrees above absolute zero, whereas everything else in the vicinity of the experiment would be very much hotter—at least several hundred degrees above absolute zero.[3] If a substantial amount of radio waves from any of these objects got into the antenna, they would utterly swamp the tiny signal from the background.

The Big Bang radiation might make up 99 per cent of the radiation flowing through the Universe, but at microwave wavelengths it was 100 million times fainter than the heat emitted by the Earth. If you had microwave glasses, you would be able to see it—well, as long as they were sensitive and could exclude the light from the ground. But it would be like trying to make out the faint uniform glow of the sky while the ground beneath you and every object around you were shining with white heat!

So Wilkinson and Roll had a formidable task ahead of them. They had to design their antenna so that when it was pointing up at the sky, as little radiation as possible found its way in from the ground and other hot objects nearby. The trumpet-shaped microwave horn was good but not quite good enough. Wilkinson and Roll supplemented it with a sort of upside-down metal skirt that surrounded the antenna. This "ground shield" made it very difficult for radiation from hot objects nearby, particularly the ground, to get into the antenna.

---

[3]Room temperature is about 300°C above absolute zero.

The cold load

But in addition to having a well-designed antenna, there was something else that was absolutely crucial to the experiment: a special device known as a "cold load." This was needed because the antenna was trying to see the coldest thing in the Universe, and no antenna could do that if it operated like a conventional radio telescope.

So how does a conventional radio telescope operate? Essentially, the radio waves picked up by a radio telescope from a star or a galaxy generate "static" in its receiver, rather like the background hiss from a radio tuned between stations. Unfortunately, lots of other things produce a similar static in the receiver. For instance, static is produced by radio waves coming from the Earth's atmosphere, and even by electrons jostling about inside both the metal of the antenna and the electronics of the receiver.

So how do astronomers tell apart the astronomical static from the spurious static? They use a simple trick. They first point their antenna at the star or distant galaxy they are interested in and note down the strength of the radio waves. Then they point the antenna at a piece of background sky nearby and take another reading. In both cases, the unwanted static created by the antenna, the receiver and the atmosphere will be the same. So if they subtract one reading from the other, they will be left with the strength of the radio waves coming from the star or galaxy. The unwanted static will have neatly cancelled out.

Of course, all that the radio astronomers will have measured is how much brighter their star or galaxy is than the background sky. But in practice the background sky will be giving out almost no radio waves so it won't matter very much.

This "on source/off source" trick works perfectly when astronomers want to look at a source of radio waves that covers only a small area of the sky—a star or distant galaxy, for instance. Then it is easy to point an antenna at a piece of background sky away from the source. But Wilkinson and Roll were planning to observe a source of radio waves that covered the entire sky. The Big Bang radiation *was* the background sky so it would be impossible to look away from it.

But if it was impossible to compare the Big Bang radiation with the background sky then it would have to be compared with something else. Wilkinson and Roll realized they would have to make an artificial source of radio waves—a "cold load." They could then point their antenna at the sky, note the strength of the radio waves, then point it at their artificial source—and take another reading. By subtracting one reading from the other they would discover how much hotter the sky was than their artificial source.

If they knew the temperature of their artificial source well then they would know the precise temperature of the Big Bang radiation. In the jargon, their artificial source of radio waves would enable them to make an "absolute" measurement: rather than simply comparing a radio source with the sky, as most radio astronomers did, they would be able to measure the true temperature of what they were looking at.

Ideally, the artificial source of radio waves should be close to the expected temperature of the Big Bang radiation—between 3 and 10 degrees above absolute zero. Wilkinson and Roll therefore decided to cool their artificial source with liquid helium, which always boils at about 4.2 degrees above absolute zero (–269°C). This is why the artificial source of radio waves was called a cold load.

Nowadays liquid helium is readily available and there is a lot of experience in handling it, but back in 1964 it was a pretty novel substance to be playing around with.[4]

It was Peter Roll who took on the task of designing and building the cold load. The important thing was to make sure that it absorbed all the radio waves that fell on it and did not reflect any back. This was because when the antenna was pointed at it, the cold load had to appear to be precisely 4.2 degrees above absolute zero. But if it reflected any radio waves at all, radio waves emitted by the metal of the antenna would bounce off the cold load straight back into the antenna. It would see the cold load plus its own reflection, causing Roll and Wilkinson to overestimate its temperature. They would assume that the cold load was at 4.2 degrees—it

[4]Liquid helium is probably the most bizarre liquid in nature. It can behave as a so-called superfluid, defying gravity by running up slopes, and squeezing through tiny holes that no other liquid could squeeze through.

was, after all, their only temperature reference—but its temperature might in fact be higher, say 6 degrees. Since they would be comparing this with the Big Bang radiation, they would underestimate its temperature, and the whole experiment could be screwed up.

It might seem a silly thing to worry about but every possible source of unwanted radio waves has to be thought about when you are attempting to measure the coldest thing in the Universe, and by definition everything in existence is hotter. "You really have to understand every detail of your instrument," says Dicke.

Roll made the cold load nonreflecting by using a length of silver-plated X-band "waveguide"—basically, just a hollow metal tube with metal rectangular cross-section. This dipped down into a vacuum flask of liquid helium. So when the antenna looked at the cold load it saw a source of radio waves at precisely 4.2 degrees.

Wilkinson and Roll arranged their instrument so that it switched from looking at the sky, then at the cold load, then back at the sky again, and this was repeated very rapidly. The electrical device that made this possible was called a "Dicke switch." Dicke, it seemed, had invented virtually everything in the field of microwave astronomy!

In fact, it was Dicke who in 1946 introduced the standard convention of measuring the brightness of a radio source in terms of an equivalent temperature. So when radio astronomers turn their telescopes on an object in the sky and say they measure a temperature of, for instance, 100 degrees above absolute zero, what they mean is that their instrument registers the same signal as it would if a body at a temperature of 100 degrees was stuck right in front of the antenna. It's just a convenience. Wilkinson and Roll expected the cosmic background radiation to be between 5 and 10 degrees above absolute zero.

A telescope like no other in the world

As Wilkinson and Roll worked on the roof of Princeton's geology building, few people walking around the campus realized that the six-foot trumpet sticking out of a pigeon coop above their heads was designed to see into the fireball at the beginning of the Universe. "Our experiment didn't attract a lot of attention on campus," admits Wilkinson. "But then we didn't go out of our way to let people know what we were doing."

From time to time, even Wilkinson and Roll thought that maybe they were just a little mad. "It wasn't obvious from the beginning that this was a good way to spend a few years," says Wilkinson. "Most people at the time believed in the Steady State theory, not the Big Bang." But at other times Wilkinson was quite optimistic about their search. "I thought we had a fifty-fifty chance of finding it," says Wilkinson.

The telescope Wilkinson and Roll were assembling had two features unique in the world—a cold load and an antenna carefully designed to reject radio waves from the ground. No other instrument in the world was capable of detecting the microwave background radiation from the Big Bang. Or so the two astronomers thought.

5

THE GHOST SIGNAL AT 4080 MEGACYCLES

Problems with an ice-cream cone antenna

In the summer of 1964, Dave Wilkinson and Peter Roll were on the brink of an epoch-making discovery. But as they busied themselves high on the roof of Princeton's geology building, assembling the radio antenna with which they intended to take the temperature of the Universe, another antenna less than an hour's drive from Princeton was already registering a peculiar and persistent hiss of radio static that was coming from every direction in the sky.

For two deeply puzzled young radio astronomers at the Bell Telephone Laboratory in Holmdel the mysterious hiss marked the beginning of the most frustrating year of their lives—a year in which they were destined to spend more time removing pigeon droppings from their antenna than actually making observations of the Universe.

Arno Penzias was 31 years old, a dynamic New Yorker who as a child had come to the USA as a refugee from Nazi Germany. Robert Wilson was a taciturn 28-year-old who had moved east after completing his graduate work at Caltech in Pasadena. In 1963, the two had teamed up to work on a very unusual radio antenna that Bell Labs had built at its Holmdel site in northern New Jersey.

The antenna, designed for satellite communications, stood on top of Crawford Hill, a low wooded knoll which barely rose above the flat monotony of the surrounding countryside. It looked nothing at all like a familiar radio dish; in fact, it was really rather hard to describe. Someone had referred to it as "an alpenhorn the size of

Robert Wilson and Arno Penzias with the 20-foot horn antenna on Crawford Hill, in New Jersey, in 1965. Despite believing at one point that the horn had picked up the microwave glow of pigeon droppings, they had in fact made one of the greatest scientific discoveries of the twentieth century. For the discovery of the afterglow of creation, Penzias and Wilson were awarded the 1978 Nobel Prize for Physics.

a boxcar," but a better description might be a giant ice-cream cone laid on its side.

A 20-foot-square opening had been cut in the side of the cone, just beneath where the ice cream should have gone. This opening collected microwaves from the sky. They were then funnelled down to a sensitive radio "receiver" installed in a cramped wooden cabin at the cone's tapered end. People could work in this cabin, fiddling with the receiver's electronics while they watched a red pen trail across a chart recorder, depicting the radio signal the antenna was picking up. The entire antenna could be turned about two separate axes so that the 20-foot opening could be pointed anywhere at all in the sky.

A giant silver beach ball in space

Despite its unusual appearance, the antenna on top of Crawford Hill was simply a standard microwave horn. In fact, it was little more than a larger version of the one Dave Wilkinson had soldered together at Princeton. Bell Labs had built it in 1960 in order to bounce radio signals off the "Echo 1" satellite, a sort of stone-age communications satellite, which was the ancestor of all the satellites that make today's world a smaller place. Echo 1 was like a silvered beach ball 100 feet in diameter. It hung in space, a brilliantly bright artificial moon in the night sky.

The problem of picking up a weak radio signal reflected from a tiny satellite was a formidable one. The engineers at Bell Labs not only had to develop an extremely sensitive radio receiver capable of measuring differences in temperature of only a few tenths of a degree, but they also had to design a very special antenna.

The basic problem they faced was that the radio signal reflected from a satellite—little more than a pinprick in the sky—would be utterly swamped by unwanted radio waves coming from nearby sources such as the ground. So the engineers at Bell Labs had to design an antenna that would keep out all radio waves except those coming from the direction of the satellite. By a coincidence, it was precisely the problem Wilkinson and Roll had to overcome in their search for the fireball radiation.

Echo 1, a sort of stone-age communications satellite, on the launchpad in 1960. The magnesium sphere on the nose cone was primed to burst open in space, releasing a silvered beach ball 100 feet in diameter. Microwaves transmitted from one antenna could be bounced off the ball and picked up by another antenna.

At Bell Labs, they solved it with the ice-cream cone design. It was almost impossible for radio waves from the ground to bend their way into the 20-foot opening when the Holmdel antenna was pointed at a source in the sky.

Echo 1 was superseded by a more sophisticated communications satellite called "Telstar" and the ice-cream cone antenna was adjusted so that it could transmit and receive microwave signals from Telstar instead. It was while the Telstar project was under way that Bell Labs hit upon the idea of hiring two radio astronomers to come and do some astronomy with their unique antenna. The company reasoned that since radio astronomers were also in the business of pushing the technology of detecting radio waves to its limits, Bell Labs might benefit from having some around.

The perfect partnership

Arno Penzias was recruited in 1962. He came straight from New York's Columbia University, where he had been a student of Charles Townes. Townes was the inventor of the maser—the microwave predecessor of the now familiar laser.[1] A year after Penzias's arrival, Bell Labs recruited Robert Wilson. At Caltech he had worked for John Bolton, an Australian who was one of the pioneers of radio astronomy.

While working on his thesis at Caltech, Wilson had got to know a Bell Labs man called Bill Jakes. Jakes showed up at Caltech at regular intervals to talk with the radio astronomers there and to ask if anyone might be interested in a job working with the 20-foot antenna at Holmdel.

Wilson had already gained a good impression of Bell Labs because he had worked with some Bell Labs people up at Caltech's Owens Valley radio observatory in northern California. The company had loaned Caltech some experimental equipment for the radio telescopes, and Wilson had helped to install it.

Shortly after he finished his thesis, Wilson applied for the job with the 20-foot horn antenna. "At the time I wasn't sure I wanted

[1]Townes was to win the Nobel Prize in 1964 for inventing the maser.

to continue in astronomy," he says. "But the Bell Labs job was an opportunity to continue doing astronomy, and I could also see a lot of other things going on at the company that were interesting to me."

Wilson got the job and arrived at Bell Labs in March 1963. He soon met Arno Penzias and they decided to join forces. "Arno and I were the only two radio astronomers at the place so it was natural for us to team up," says Wilson.

It was destined to be a perfect partnership. Not only did the two radio astronomers have complementary technical skills, but their personalities complemented each other as well. Wilson was quiet and cautious while Penzias was brash and outspoken. But though on the surface they were like chalk and cheese, they shared one important characteristic that would go a long way to ensuring their eventual success. Both men were meticulous and painstakingly thorough when it came to doing science.

The halo around the Galaxy

Although Arno Penzias had been at Bell Labs for a year, he had not yet been able to get his hands on the 20-foot horn because it was still being used for Telstar. But that all changed shortly after he and Wilson teamed up. "The Telstar people finally agreed we could do some astronomy with it," says Wilson.

Immediately, Penzias and Wilson set about modifying the antenna for astronomy. "The 20-foot antenna was unique," says Wilson. "It produced very little in the way of unwanted radio signals, and there was the possibility of determining exactly how big those signals were and where they were coming from."

This made the antenna good for making "absolute measurements"—that is, measuring how bright at source a radio wavelength really was rather than simply comparing it with the sky background. Of course, if Penzias and Wilson were to exploit it for this purpose they would need an artificial source of radio waves with which to compare any astronomical source.

So it was that Penzias began building a cold load. The device he came to build was remarkably similar to the one Peter Roll was building at Princeton. Both devices used a piece of waveguide and both were cooled by liquid helium to just 4 degrees above absolute zero. A crucial element of the design of the Bell Labs experiment was a switch that allowed the temperature of the sky and the cold load to be compared rapidly.

In 1964, it was probably fair to say that there were only two liquid helium cold loads in existence in the world. It was rather a coincidence that they had been built independently by two groups of astronomers who were unaware of each other's existence and yet were only thirty miles apart.

Now equipped with its cold load, the Holmdel antenna was ideally suited for picking up a faint background signal from the sky. And this is precisely what Penzias and Wilson intended to do with it.

For his thesis with John Bolton at Caltech, Wilson had made a map of the Milky Way at radio wavelengths. He had suspected that surrounding the starry disc of the Milky Way was a vast "halo" of gas glowing faintly at radio wavelengths but he had been unable to prove it. The reason was that he had made his map by using the standard technique of comparing the brightness of the Milky Way with the background sky. His technique was therefore incapable of measuring the brightness of the Milky Way's faintly glowing halo, as in effect this was the background sky.

The 20-foot horn, with its ability to measure the radio signals from faint background regions of the sky, was an ideal instrument for measuring the weak radiation from the Milky Way's halo. Penzias and Wilson decided to look at a wavelength of 21 centimeters. If the halo glowed at all, it would glow at 21 centimeters. This was because it should be made of neutral hydrogen gas, which broadcasts a very distinct radio signature at this wavelength.

But the two radio astronomers knew that actually looking for the galactic halo at 21 centimeters would be tough. The halo was likely to be very faint, and to register no more than a temperature of one degree at their antenna. Other, unwanted, signals from the antenna and receiver and the atmosphere would be much larger. So it was clear to Penzias and Wilson that before they attempted the halo measurement they would really need to understand their instrument and know where all the unwanted radio signals were coming from and just how big they were.

The ghost signal at 4080 megahertz

The Telstar people had left the 20-foot horn with a receiver set up for a wavelength of 7.35 centimeters, or 4080 megacycles per second. Penzias and Wilson therefore decided to take advantage of this and try to understand completely what was happening in their instrument at this wavelength before going to the trouble of building another receiver sensitive to 21 centimeters.

It turned out that observing the sky at 7.35 centimeters would be a particularly neat test of their ability to measure the temperature of the sky because at this wavelength the Milky Way's halo should be essentially invisible. So when the antenna was turned on the sky and all the sources of unwanted static were accounted for, the 20-foot antenna should register only a signal from the antenna structure itself and this should be almost zero. So if Penzias and Wilson pointed their antenna at the sky and had no signal left when they had done their accounting then all would be fine.

Penzias and Wilson did just this in June 1964. They fully expected to measure a sky temperature of zero degrees. But it was immediately clear that something was very wrong. Their horn was generating more radio static than they expected. Even when they had accounted for every source of unwanted radio waves the instrument was still registering a signal. It was precisely what would be produced by a body at a temperature of just 3.5 degrees above absolute zero.

"When we made that measurement, Arno's first reaction was 'Well, I made a good cold load,' " says Wilson. If the cold load had been reflecting any radio waves back into the antenna then the cold load would have appeared to be hotter than 4.2 degrees and so screwed up Penzias and Wilson's accounting job.

After satisfying themselves that the cold load was fine, the two astronomers wondered whether they might be picking up a man-made signal in the urban environment of northern New Jersey. "The best place to do radio astronomy is a completely isolated valley that's shielded from all radio interference," says Wilson. "But the Holmdel antenna had been built on top of a hill so that it would get complete coverage of the sky for satellite communications."

If the anomalous signal was man-made then the obvious source was New York City, thirty miles to the north. But when Pen-

Edward Ohm (right) at the 20-foot horn antenna at Holmdel, New Jersey. He had noticed as early as 1961 that the horn was registering more static than expected when pointed at the sky but had not realized its significance.

zias and Wilson checked this out by pointing their antenna in the direction of the City, the signal on their chart recorder did not jump. In fact, the ghost signal at 4080 megacycles stayed the same wherever they pointed the instrument around the horizon.

A persistent problem

It turned out that Penzias and Wilson were not the first people to encounter the problem of the peculiar excess signal. As early as 1961, Ed Ohm, an engineer working on the 20-foot horn, had noticed that the instrument was registering more static than expected when it was pointed at the sky. With the horn receiving signals bounced off the Echo satellite, Ohm had added up all the sources of unwanted radio static. He had found that the antenna was picking up something like three degrees more than he could account for.

Ohm did not pay too much attention to this excess temperature because some of the contributions in his sum were only uncertain by more than three degrees. Without a cold load it was impossible to pin down where the excess static was coming from. Nonetheless, Ohm published his result in the *Bell System Technical Journal*.

Penzias and Wilson wondered whether their "amplifier" circuits were producing the excess signal. Amplifiers are part of any radio receiver. They are needed because the electrical currents generated by radio waves in an antenna are so tiny that practical detectors usually cannot register them. Instead, the currents have to be magnified electronically by "amplifiers" before they reach a detector.

The two astronomers compared the signal coming from the antenna when it was looking at the cold load and when it was not. Since the signal from the amplifier circuits had to be the same in both cases, it neatly cancelled out. What was left was the signal coming from the Holmdel antenna alone. They knew this was made up of contributions from the metal structure of the antenna, from the Earth's atmosphere and any astronomical sources of radio waves that happened to be in the direction the antenna was pointing.

The static from the atmosphere was easy to identify and subtract because of its distinctive characteristic: the hiss was strongest when the antenna was pointed at the horizon—the direction in which the atmosphere is thickest—and weakest when it was pointed straight up—where the atmosphere is at its thinnest.

Of course, the anomalous radio signal could have been real. But this seemed too ridiculous to contemplate. For a start, it could not be coming from the Sun or the Milky Way because neither covered the whole sky, and the signal quite definitely did. The only other possibility was that the signal was coming from the Universe as a whole. But these astronomers knew of no astronomical source that could be generating such a constant radio signal. Clearly, there must be a fault in the antenna causing it to generate more static than Penzias and Wilson had realized. They were confident that the antenna was generating very little static. This was one characteristic of the Holmdel instrument that had convinced them in the first place that it was uniquely suited for making the difficult galactic halo measurement. But Penzias and Wilson were nothing if not meticulous. They decided to look at the antenna in detail.

They died for science!

Their gaze settled on a pair of pigeons that were roosting deep inside the ice-cream cone antenna, just at the point where it entered the wooden cabin. "It was a nice comfy place in there because the end of the antenna was up in our heated control room," says Wilson.

It might have been warm in there but it was difficult to build a nest. Every few days, Penzias and Wilson turned their antenna, tipping the pigeons on their heads.

The pigeons had left their distinctive mark on the inside of the great ice-cream cone. To Penzias, a radio engineer through and through, it was "a white dielectric material." But to anyone else it was simply pigeon shit!

"Until now, we'd been operating quite happily with the stuff in place," says Wilson. "There were no big heaps of the stuff because anything loose fell off whenever we turned the antenna around."

So could the pigeon droppings that were coating the inside of the antenna be responsible for the mysterious static? Since everything above absolute zero gives out radio waves, the pigeon droppings would certainly be glowing at microwave wavelengths. By now, Penzias and Wilson were desperate enough to consider anything.

They decided to eject the pigeons. But this proved to be no easy task. From a local hardware store they bought a "Hav-A-Heart" trap, which they put at the end of the antenna after removing some of their receiver.

The Hav-A-Heart trap was a wire mesh cylinder with a gate at either end. You put food on a feeding tray at the center of the cylinder and in theory when an animal or bird walked in and disturbed the feeding tray the two gates were triggered to fall. It worked perfectly. "I think we got one pigeon one day and the other the next," says Wilson.

They put the pair of pigeons in a box and mailed them to Whippany, another New Jersey site of Bell Labs forty miles to the northwest of Holmdel. "We sent them there because it was the most distant place we could send them in the company mail," says Wilson. At Whippany, a man had agreed to accept them and turn them loose.

Once the pigeons were on their way, Penzias and Wilson set about removing the pigeon shit. They climbed into the gloomy interior of the horn antenna armed with brooms. "It wasn't a big job," says Wilson. "After an hour of sweeping, we'd removed everything."

Penzias and Wilson thought they had seen the last of their pigeons but they were wrong. "Two days later, the pigeons were back in the antenna," says Wilson.

"By now we decided we had given them a good chance," says Wilson. "There was a guy in the machine shop who was a pigeon fancier and he told us these were junk pigeons, and we were not to worry about them. One day, he brought in his shot gun and blew them to kingdom come."

"The oddest thing," says Wilson, "is that since our pair of pigeons, no others have ever nested up there in the 20-foot horn."

With the pigeons gone for good, Penzias and Wilson thoroughly cleaned the interior of their antenna. The antenna was made of aluminum sheets that were riveted to aluminum beams. Thinking that the rivets might be causing the spurious signal, they put aluminum tape over them. So careful were they that they had even checked that the adhesive on the back of the tape generated negligible radio waves. Now, surely, they had thought of everything. At long last they would be able to do some radio astronomy.

Penzias and Wilson pointed the 20-foot antenna at the sky and looked at the reading on their chart recorder. To their dismay they saw that the spurious static had decreased only slightly. It had not gone away. The horn was still registering an anomalous temperature of 3.5 degrees above absolute zero.

By now the excess signal had persisted for almost a year. As far as Penzias and Wilson could tell, it was the same in all directions and it did not vary with the seasons. They were also able to rule out two additional sources of radio waves. It could not be in the Solar System because any source should have moved around the sky as the Earth orbited the Sun. Also, it could not be due to a nuclear test. In 1962, a high-altitude nuclear explosion had injected ionized particles into the Van Allen radiation belts, high above the earth. But any radiation from this source should have reduced considerably within a year of the explosion.

Penzias and Wilson were at a loss for any further explanation. This tiny but persistent effect had sabotaged their plan to observe the halo of the Galaxy. But just when they were at their wits' end, Penzias happened to make a phone call . . .

6

A TALE OF TWO TELEPHONE CALLS

How the fireball radiation came to be found

It is a fair claim that one of the greatest scientific discoveries of the twentieth century was made by telephone. In fact, not by one telephone call but by two.

In April 1965, Arno Penzias phoned Bernie Burke, a prominent American radio astronomer at the Carnegie Institution's Department of Terrestrial Magnetism in Washington DC. Penzias's call was prompted not by the problem with the 20-foot antenna but by another matter altogether, and he would never have mentioned the irritating static had Burke not asked him in passing how the experiment on Crawford Hill was going. Immediately, Penzias launched into a long complaint about the irritating signal that would not go away and about how frustrating it was trying to track down its source.

Burke sat up. One of his colleagues, Ken Turner, had told him about a search that was under way for just such a signal at Princeton. Could that be what Penzias and Wilson had picked up?

He tried to recall what Turner had told him. Turner had been to a talk the previous month given by Jim Peebles, a friend from his days as a graduate student at Princeton (Turner's supervisor had been none other than Bob Dicke). The talk Peebles had given was at a meeting of the American Physical Society held at Columbia University. As far as Burke could remember from what Turner had told him, it was about fireball radiation being an unavoidable consequence of a hot Big Bang. Peebles had argued that if the Universe's helium was indeed produced in the Big Bang, then today's Universe should be filled with microwaves with a temperature of less than 10 degrees above absolute zero. This tepid afterglow of creation was detectable with current technology. In fact, Dicke's group at Princeton had already embarked on a search for it.

Burke immediately alerted Penzias to the possibility that the anomalous signal might be the leftover glimmer of the Big Bang. It was music to Penzias's ears. By now he was desperate to find an explanation—any explanation—for the 3.5 degree excess temperature. He got on the phone to Dicke.

Well boys, we've been scooped!

When the phone rang in Dicke's office at Princeton, Dicke had company. Seated in a circle around his desk, sipping cups of coffee and eating sandwiches, were his three disciples—Wilkinson, Roll and Peebles. "Every week we used to have these brown-bag lunches to chat about how our experiment was going and talk about what we ought to be doing next," says Wilkinson. "Arno's call came during one of those gatherings."

Dicke's telephone conversation was rather one-sided. He mostly listened, now and then nodding and repeating phrases familiar to the others in the office. Wilkinson's ears pricked up the moment he heard Dicke mutter the words "horn antenna."

Peebles remembers the conversation vividly. "I seem to recall it involved such mysterious things as pigeon droppings," he says.

Nobody in the Princeton group knew Arno Penzias or Robert Wilson but the team was well aware of the 20-foot antenna Bell Labs had built out at Holmdel for the Echo project. Roll and Wilkinson had learned about it while scouring the microwave journals before starting on their experiment. "It was abundantly clear to us that Bell Labs had the best antenna around," says Wilkinson.

Roll and Wilkinson had come across Ed Ohm's papers in the *Bell System Technical Journal*, and carefully read them. They had concluded that there were clear signs the 20-foot antenna was picking up something unusual from all over the sky. But Ohm did not have a cold load. Without that, there was no way to tell whether he was really seeing the cosmic background radiation or simply a spurious radio signal from a more mundane source.

On the telephone, Dicke continued to repeat familiar microwave phrases. Then suddenly he said, "cold load."

"As soon as we heard those words, we knew the game was up," says Wilkinson.

Moments later Dicke hung up the phone. He turned to Peebles, Roll and Wilkinson. "Well boys," he said, "we've been scooped!"

Two first-class astronomers

The next day, Dicke, Roll and Wilkinson drove the thirty miles over to Holmdel to take a look at the Bell Labs apparatus. They were met by Penzias and Wilson on Crawford Hill.

Although the two groups of astronomers had never met, Penzias and Wilson knew the name of Bob Dicke. "I was considerably in awe of him," says Wilson. "He was the grand old man of microwaves."

Once the introductions were over, Penzias and Wilson led their visitors over to the 20-foot antenna and began to show them the equipment. "I don't remember them being unusually inquisitive," says Wilson.

If they were not very inquisitive, it was because Dicke had already asked most of the pertinent questions on the telephone the day before. "Before we went over to Bell Labs, we were pretty convinced that they'd found the Big Bang radiation," says Wilkinson. "You see, the experiment to look for it is a rather simple one if you have the right apparatus. There are half a dozen things that you have to do right, then the temperature of the background just pops out."

Another reason that Dicke's group asked so few questions was that they already knew most of the answers. The Bell Labs apparatus turned out to be remarkably similar to the experiment Roll and Wilkinson were building back at Princeton. In particular, Penzias's helium cold load was almost identical to the one Peter Roll had designed. "The similarities meant we caught on very quickly," says Wilkinson.

Dicke's group was quickly convinced that Penzias and Wilson were first-rate radio astronomers. "I was really impressed that they had hung in there on a problem that wasn't central to what they were doing," says Wilkinson. "Here was this thing that they really wanted to understand. And they'd been working on it for a year, worrying at it and never letting go. They had taken enormous care to rule out the more obvious explanations for their puzzling signal."

What worried the Princeton team most was that unwanted radio waves from the ground might be somehow finding their way into the 20-foot Holmdel antenna. "It was impossible to shield the horn from the ground because it was such a big thing," says Wilkinson. But Penzias and Wilson were able to convince their visitors that when the Holmdel antenna was pointed at the sky, very little ground radiation could bend its way into the 20-foot opening of the ice-cream cone antenna.

Wilkinson and the others pored over Penzias and Wilson's data—wiggly red lines on chart recorders. By now, they were satisfied with what they had seen. "Penzias and Wilson were looking at a wavelength where there shouldn't have been any signal at all so we were convinced they must be seeing the primeval fireball," says Wilkinson.

The effect they had measured was small—no more than a few degrees. Any other instrument in the world would have missed it, but the Holmdel antenna was uniquely suited for distinguishing a weak background signal from other, much stronger sources. There on the chart recorder was a cryptic message from the very beginning of time.

Message from the beginning of time

If they were right, it was the most important discovery in cosmology since Edwin Hubble had found that the Universe was expanding back in 1929. Permeating every pore of the Universe was a tepid radiation, the "afterglow" of the titanic fireball in which the Universe was born. Before the Holmdel antenna had intercepted it, the radiation had been streaming across empty space for an incredible 15 billion years. Penzias and Wilson had stumbled on the oldest "fossil" in creation, carrying with it an imprint of the Universe as it was soon after the creation event itself.

The temperature of the background radiation was the temperature of the early Universe, greatly reduced by the enormous expansion it had undergone since.

When the radiation broke free of matter, the Universe was at a temperature of about 3,000 degrees.[1] But while it had been flying to us across space, the Universe had expanded a thousand times in size, diluting the temperature of the radiation by exactly the same amount so that today it appeared to be only about 3 degrees above absolute zero.

The temperature of about 3 degrees above absolute zero *is* the temperature of the Universe. Although the stars are very hot and very numerous, when their temperatures are averaged over all of space, their contribution to the temperature of the Universe is completely negligible compared with the fireball radiation.

The cosmic background radiation came from the time when it first became cool enough for atoms to form. At this instant, about 300,000 years after the Big Bang, the rapidly cooling fireball suddenly became transparent to light. Photons which had bounced from particle to particle in the fog of the fireball were suddenly able to move freely. And they have been doing so ever since, gradually losing energy as the Universe has grown in size.

It may seem peculiar that the cosmic background radiation is arriving at the Earth only today, 15 billion years after the Big Bang. After all, in a sense we were in the Big Bang (or at least, the particles of matter that would one day condense to form the Earth were in the Big Bang) and the fireball radiation was all around us. Surely it should have already passed us by now?

Well, radiation which in the Big Bang was emitted by matter in our immediate neighborhood has already passed us. Forgetting for a moment that the Universe has expanded a lot since the Big Bang,

[1]The temperature dropping to about 3,000 degrees also signalled another significant event—the point at which the energy density of the radiation, or photons, in the Universe fell below that of matter. From then on, the Universe was dominated by matter and by the force of gravity acting on that matter.

it is true to say that radiation emitted 15 billion light years[2] from us is just arriving at the Earth today. On the other hand, radiation that was emitted 10 billion light years away would have arrived 10 billion years after the Big Bang—or just as the Sun and Earth were forming 5 billion years ago.

The expansion of the Universe complicates matters a little because when those photons of the Big Bang radiation arriving at the Earth today broke free of matter, the Universe was only a thousandth of its present size. The photons have therefore taken 15 billion years to cross a gap that was originally only 15 million light years wide. It is as though you were trying to sprint 100 meters on a track that has grown a thousand times longer while you are running.

The detection of the cosmic background radiation by Penzias and Wilson meant that the Big Bang theory was triumphant. If Martin Ryle's work at Cambridge on radio galaxies had sent the Steady State theory reeling, the discovery of the afterglow of creation dealt it a knock-out blow.

For the second time in its history, scientists at Bell Labs in Holmdel had made a great scientific discovery serendipitously. Back in 1931, a 26-year-old Bell Labs physicist named Carl Jansky, who had been investigating possible sources of radio interference, detected a weak static that seemed to be coming from the Milky Way, and thus invented the science of radio astronomy.

The first of many wonderful things

By rights Dicke's group should have been sick that they had been scooped. But if they were, it was not the impression Wilson got. "I don't remember them appearing deflated," says Wilson. "That didn't come across strongly at all."

"At the time, it didn't bother me that we had been scooped," says Wilkinson. "Peter and I were too busy getting our experiment going to worry. Also, I was young. I thought this was just one of a

---

[2]A light year is the distance light travels in a year.

series of wonderful things that was going to happen to me in my career. But, of course, discoveries like this come along only every decade or so."

Ironically, it had never occurred to Wilkinson and Roll to ask Bell Labs to use the 20-foot antenna, despite the fact they recognized it as the only instrument in the world that could detect the fireball radiation. "If they had come and asked, I'm sure Bell Labs would have given them permission," says Wilson. "Arno and I would have been left standing watching on the sidelines."

Peebles remembers Dicke and the others coming back from Bell Labs and pronouncing themselves impressed by what they had seen. "I don't remember feeling particularly excited by the discovery nor deeply disappointed that it had not been a Princeton discovery," he says. "You see it was by no means obvious that this was radiation from the Big Bang. It could still have turned out to be something quite mundane."

At last we can do some real science!

Penzias and Wilson were both slow to accept the cosmological origin of their mysterious signal. "They'd spent so long focusing on all the mundane explanations—like pigeon droppings," says Peebles, "that I think it took them a while to realize just how great a discovery they had really made."

In fact, it was at least a year before the two astronomers would accept that their anomalous signal came from the Big Bang. "We had made a measurement which we thought would hold up," says Wilson. "But we weren't so sure that the cosmology would."

Wilson had another reason for dragging his feet. "I'd rather liked the Steady State theory," he says. Inadvertently, he had helped to destroy it.

But though Penzias and Wilson were a bit dubious about the Big Bang idea, both were very pleased to finally have an explanation for the problem that had been troubling them for so long. "When we came along, they were at a complete loss for any other explanations," says Peebles. "They were feeling driven against a wall."

"They desperately wanted to use the antenna to do some radio astronomy," says Wilkinson.

This is certainly illustrated by Penzias's immediate reaction to the Princeton explanation. According to Peebles, in one of their early telephone conversations, Penzias said: "Well, that's a big relief. We understand this thing at last. Now we can forget it and go and do some real science!" But rarely had there been a scientific result that was less likely to be forgotten!

The parallels with the century's other great cosmological discovery were striking. Both the expansion of the Universe and the fireball radiation had been found by scientists who were completely unaware that predictions of the phenomenon had been made many years before in the scientific literature. Which makes you wonder whether scientists ever read the scientific literature at all—or remember what they read!

The world learns of the discovery

The Princeton and Bell Labs groups decided to announce the discovery in two scientific papers, published side by side, in *Astrophysical Journal Letters.*

Two weeks before the papers were due to appear in print, Wilson finally began to realize how important a discovery he and Penzias had made. The phone rang out at Crawford Hill, and on the other end was Walter Sullivan, the science reporter of *The New York Times.*

Sullivan had been on the trail of another story entirely when he had happened to call the offices of *Astrophysical Journal.* "For some unknown reason they leaked our paper to him," says Wilson. Sullivan grilled Penzias about the work with the 20-foot antenna.

At the time of the phone call, Wilson's father was visiting him from Texas. An habitual early riser, the next day he got up well before his son to walk down to the local drugstore. When he came back, he had a copy of *The New York Times.* He thrust it in the face of his bleary-eyed son. There on the front page was a picture of the 20-foot horn with a description of the *Astrophysical Journal* pa-

"All the News That's Fit to Print"

The New York

VOL. CXIV..No. 39,199. © 1965 by The New York Times Company, Times Square, New York, N. Y. 10036 NEW YORK, FRIDAY, MAY 21, 1965.

Signals Imply a 'Big Bang' Universe

Horn antenna, used in space exploration, at the Bell Laboratories in Holmdel, N. J.

By WALTER SULLIVAN

Scientists at the Bell Telephone Laboratories have observed what a group at Princeton University believes may be remnants of an explosion that gave birth to the universe.

These remnants are thought to have originated in the burst of light from that cataclysmic event.

Such a primordial explosion is embodied in the "big bang" theory of the universe. It seeks to explain the observation that virtually all distant galaxies are flying away from the earth. Their motion implies that they all originated at a single point 10 or 15 billion years ago.

The Bell observations, made by Drs. Arno A. Penzias and Robert W. Wilson from a hilltop in Holmdel, N. J., were of radio waves that appear to be flying in all directions through the universe. Since radio waves and light waves are identical, except for their wavelength, these are thought to be remnants of light waves from the primordial flash.

The waves were stretched into radio waves by the vast expansion of the universe that has occurred since the explosion and release of the waves from the expanding gas cloud born of the fireball.

In what may prove to be one of the most remarkable coincidences in scientific history, the existence of such waves was predicted at

Continued on Page 18, Column 1

NDSAY PROPOSES DDING CIVILIANS O POLICE BOARD

Vould Supplement 3 From epartment Now on Panel to Review Complaints

E PATROLMAN URGED

izen Members Preferred 'ith Judicial Training— Rights Groups Critical

xt of Lindsay's speech is printed on Page 12.

By RICHARD WITKIN

epresentative John V. Lind-, in the first major speech of mayoral campaign, proposed night the addition of four lians to the Police Depart-it's three-man board for re-ving complaints of police tality.

he Manhattan Republican gested that the four new nbers be picked by the May-rom a list submitted to him a committee composed of n and women of unquestion-stature."

inal responsibility for any iplinary action against ac-d policemen would rest with Police Commissioner. This standard with virtually all lian-review plans.

Middle Road Sought

r. Lindsay said "it is time a change" in the sensitive of relationships between cemen and average citizens. It is time for us to exchange ect rather than to exchange ilts," he added. "But this not happen until all of the ple of our city feel and be-

BEAME'S DEMAN FOR AUDIT AT FAI UPHELD BY COUR

Appeals Bench Unanimous Backs City Controller— Pickets Win Case

By R. W. APPLE Jr.
Special to The New York Times

ALBANY, May 20—The ri of City Controller Abraham Beame to look into the financ records of the World's Fair v unanimously affirmed today the Court of Appeals.

The court issued no opini It simply filed an order holding the decisions of t lower courts, one of which described the Controller's mand for access to the fa books as a "fitting and prop performance of his duties.

Mr. Beame said in Manhat that he would send audit from his office to the fa headquarters at Flushing Me ow, Queens, as soon as p sible.

The Court of Appeals, highest tribunal in the st held also today that four m bers of the Congress of Ra Equality had violated no l by picketing the Flor Pavilion at the fair on A 28 of last year.

Dissenting Opinion

In this case, the court s 5 to 2. Associate Justices J Van Voorhis and John Scileppi, the dissenters, arg that the pickets were guilty both disorderly conduct illegal trespassing on the fa private property.

[Regarding the fair's fina cial obligations, a spokesm for the fair said yesterd that a document it issued la year on priorities for a funds remaining when the e hibition closed had no le

KENNEDY CRITICAL OF PARTY IN STATE

Says Democratic Legislators Reneged on Redistricting

Jersey Central Asks $5.6 Million Subsidy For Commuter Runs

By ALFRED E. CLARK
Special to The New York Times

NEWARK, May 20—The president of the Jersey Central

PLAN IS SPEEDED ON NEW AIRLINER

Johnson's Order on Building Supersonic Test Plane Is

The New York Times *announces the discovery of the cosmic background radiation on 21 May 1965.*

per. "For the first time, I really got the impression the world was taking this thing seriously," says Wilson.

George Gamow, by now retired, read the story in *The New York Times*. To his dismay, he saw no mention of his name, nor those of Ralph Alpher or Robert Herman. It is fair to say that he awaited the publication of the scientific papers with intense interest.

The papers duly came out. The title of Penzias and Wilson's gave nothing away: "A Measurement of Excess Antenna Temperature at 4800 Megacycles per Second." Rarely can such an important scientific discovery have been disguised so well!

In the paper, the two Bell Labs astronomers wrote: "Measurements of the effective zenith noise temperature of the 20-foot horn-reflector antenna at the Crawford Hill Laboratory, Holmdel, New Jersey, at 4080 megacycles per second have yielded a value of about 3.5 degrees higher than expected."

And that was basically all Penzias and Wilson said. Nowhere in their brief paper did they mention that the radiation they had picked up might have come straight from a hot Big Bang. They merely noted: "A possible explanation for the observed excess noise temperature is the one by Dicke, Peebles, Roll and Wilkinson in the companion letter in this issue."

"I think they were rather overcautious," says Wilkinson.

"Their paper was written in such a way that it could have been almost anything they'd found," says Dicke.

"In contrast, our group really went out on a limb," says Wilkinson. "In our paper, we were interpreting a single microwave measurement as proof of the existence of the Big Bang radiation."

"In fact, Penzias and Wilson weren't even going to write a paper at all until we told them we were writing one," says Dicke.

Wilson says the reason he and Penzias did not write about the Big Bang theory of the origin of the background radiation was because they were not involved in that work. "We also thought that our measurement was independent of the theory and might outlive it," he says.

"We were pleased that the mysterious noise appearing in our antenna had an explanation of any kind, especially one with such cosmological implications. Our mood, however, remained one of cautious optimism for some time."

The Gamow controversy

The moment the two scientific papers were published, Gamow made a beeline for his library. He raced through the two papers, becoming increasingly angry. Nowhere was there a mention of his ground-breaking work in the 1940s. Gamow, Alpher and Herman had not only published the results of their hot Big Bang calculations in a series of technical articles in *The Physical Review* but they had written numerous popular accounts of their work as well. For instance, in 1952 Gamow published a book for lay readers called *The Creation of the Universe* in which he talked about the cooking of helium in a hot Big Bang and how this was connected to the temperature of the Universe. Four years later, Gamow aired his ideas in an article in the popular magazine *Scientific American*.

But all these accounts were missed entirely by Dicke's team at Princeton. "We absolutely didn't know about Gamow's work," says Wilkinson. "When Jim Peebles and I were searching through the scientific literature to see what had already been done, we read only the microwave journals so we never saw any of Gamow's stuff."

One of the problems was that before Penzias and Wilson's discovery of the cosmic background radiation, cosmology was not really a distinct field. "There was no cosmology literature," says Wilkinson. "The scientific papers that were published—and there were not many—were published all over the place. Even today, thirty years later, I'm still finding papers on the cosmic background radiation that I never knew existed."

But though it is easy to understand how Wilkinson and Peebles missed Gamow's work, it is harder to explain how Dicke could have missed it. Several years earlier he had attended a talk Gamow had given at Princeton about making elements in a hot Big Bang. "Gamow spoke about a Universe in which you start with a mass of cold neutrons which suddenly explode in a Big Bang," he says. "But that's all I can recall about what he said."

And the connection between Dicke and Gamow does not end here. It turns out that the very same issue of *The Physical Review* that contained George Gamow's first 1940s paper on the hot Big Bang also contained a paper by Dicke. That might not seem too much of a coincidence but, in a throwaway remark in his own paper, Dicke actually made a comment about the possibility of a microwave background radiation in the Universe.

An attack of amnesia

As part of his wartime radar work, Dicke and his colleagues had gone to Florida to measure the radio waves coming from water vapor in the moist atmosphere. As an aside, he had wondered whether the sky might be glowing uniformly with microwaves. If such a uniform glow existed, it would have to be coming from the Universe as a whole since nearby sources, such as a planet or the Milky Way, would fill only a small part of the sky.

Dicke concluded that there was no such sky-glow that he could measure. In fact, he put it more precisely in his paper in *The Physical Review*, stating that the temperature of any "radiation coming from cosmic matter" had to be less than twenty degrees above absolute zero.[3]

Dicke had thus attempted the first ever measurement of the Universe's radiation background. But ironically he had forgotten all about it, and so, too, had everyone else. "Jim stumbled on it only when we were reading through the microwave literature," says Wilkinson. In the cosmic background field not only did people often overlook each other's work, they sometimes even overlooked their own!

But such forgetfulness was hardly likely to console Gamow, Alpher and Herman. The irony was that the last thing anyone wanted to do was upset Gamow. He was an idol to the young radio astronomers at Princeton and Bell Labs.

"Gamow was one of my heroes," says Wilkinson. "I read all of his popular books in high school. He was probably the reason I got into science in the first place." Robert Wilson had also been turned onto science by reading Gamow's popular books.

All of them realized that Gamow was one of the most intuitive and inventive physicists of the twentieth century. "He had the

---

[3]The technique Dicke used and the receivers available in the 1940s were not capable of detecting a uniform background as cold as three degrees above absolute zero.

ability to ferret out the essential elements of the most complicated physics," says Peebles. "It was that ability he used to effect when tackling the problem of the Big Bang and the fireball radiation."

Peebles and the rest felt guilty that they had not given due credit to Gamow's group. "We simply did not do our homework," he says. "We should have gone through the literature and got every possible reference to this thing. In fact, it was a couple of years before we did that." This failure to right the wrong immediately ensured that Gamow, Alpher and Herman would remain bitter about the way they had been treated.

"I tried to do all I could to bring Gamow into the whole story as much as possible," says Wilkinson. Soon after the momentous events of spring 1965, he and Peebles decided to write an article about the discovery for the magazine *Physics Today*. Before putting pen to paper, they went back and read the papers of Gamow, Alpher and Herman. But the article never got past the rough draft stage. "Alpher and Herman took issue with our version," says Wilkinson. "They wrote us a rather strong letter. So in the end we withdrew the article and never published it."

Perhaps, if someone in the Princeton team had actually telephoned Gamow at the outset, and asked him just what his group had done and when, then all the misunderstandings would have been avoided.

Arno Penzias tried his best to smooth things over with Gamow but feelings were simply running too high. "I don't think Gamow ever really forgave Dicke and his group," says Wilson. "As for us, I don't know exactly how he felt."

Gamow remained bitter until his death, in 1968, just three years after the definitive proof of the hot Big Bang he had championed. "Alpher and Herman never got over it completely either," says Wilson.

Twin injustices

Alpher and Herman perhaps had reason, for they suffered twin injustices. In the beginning, neither they nor Gamow were credited for their work on the hot Big Bang. But later, when people did give credit, they often cited Gamow alone for predicting the cosmic background radiation. This was particularly galling since this was one consequence of the primordial fireball which he had overlooked and which Alpher and Herman had presented on their own in *Nature* in 1948.

The controversy was not helped by Gamow himself, who could be rather cavalier in giving due credit and who failed in several of his later scientific papers to mention Alpher and Herman when he discussed the fireball radiation. So when Alpher and Herman got upset at someone for wrongly crediting Gamow with their work, Gamow was often the guilty party in the first place for sowing seeds of confusion in the scientific literature.

Alpher and Herman speculated a lot about why they were overlooked by the astrophysicists. They thought that the fact that they were outsiders may have had something to do with it. Both spent a considerable part of their scientific careers in industry. Alpher worked at General Electric between 1955 and 1986, and Herman at General Motors from 1956 to 1979. These were precisely the years when cosmology came of age as a science and first caught the attention of the public at large.

Whatever the reasons for being overlooked, nowadays the history books give Alpher and Herman their rightful place. The wounds seem to be healing at last. "These days they will even come to cosmology meetings and talk about it," says Wilson.

7

AFTERGLOW OF CREATION

Why did nobody find the fireball radiation earlier?

The discovery of the cosmic microwave background provides a wonderful example of the way science is really done. Though the writers of textbooks—and very often scientists themselves—would like us to believe that science progresses in a series of logical steps, taken coolly and calmly, one after another, this is patently not so. Far from being orderly, the progress of science is more like that of a drunkard staggering two steps backward for every three in the forward direction, and making the odd sideways lurch just for good measure. Consider again the story of how the fireball radiation came to light . . .

In the late 1940s, George Gamow and his co-workers guessed that if the Universe had begun in a Big Bang, the early Universe would have been filled with intense radiation, and that the pale afterglow should still be around 15 billion years later. But though they investigated the possibility of looking for the fireball radiation, they were told by radio astronomers that it was undetectable. Everyone forgot about the relic radiation because Gamow's theory was discredited.

But a decade and a half later Bob Dicke rediscovered the fireball radiation—for an entirely different reason. He decided that a search for the pale afterglow of creation was feasible and gave two young radio astronomers the job of looking for it. But on the eve of their attempt (and here the story descends into farce), another pair of astronomers stumbled on the cosmic background radiation entirely by accident, after first thinking they might be seeing the faint radio-glow of pigeon droppings.

Why was the fireball radiation not discovered earlier?

It's a strange tale—and it gets stranger. Consider the baffling question of why the cosmic microwave background was not discovered earlier. After all, it had been predicted a full seventeen years before that fateful phone call from Arno Penzias at Bell Labs to Bob Dicke at Princeton.

The question has long puzzled Dave Wilkinson. "I've often wondered why nobody in that seventeen years put two and two together," he says. "Not only was the microwave radiometer a standard instrument in radio astronomy, but Gamow's group had publicized the idea that there ought to be microwave radiation in the Universe with a temperature of just a few degrees. Gamow had even written popular articles about it in *Scientific American*. To go and look for the radiation, all you needed were two things—a good microwave horn and a cold load."

Robert Wilson also thinks it was amazing that no one carried out a search for the relic radiation earlier. "At any time after Alpher and Herman made their prediction, it could have been checked," he says. "If Bob Dicke had decided to look for the fireball radiation, he could have done it with World War Two equipment. In fact, a radio receiver like Wilkinson's could probably have been built not too long after the war.

"Of course, Alpher and Herman did go and talk to some radio astronomers who sort of said, no, the measurement is impossible. But I'm sure that if Dicke had thought of doing it, he would have done it and succeeded."

The reason Dicke did not do it was because it simply never occurred to him, something that he kicks himself for today. "On a number of occasions during and after the war, I could have used my microwave receiver to do some interesting astronomy," says Dicke. "But I missed them all. I was kind of stupid. You see, at the time I didn't quite realize what astronomy was. I'd only ever done one course on the subject."

But not everyone overlooked the prediction of Alpher and Herman. In the Soviet Union, a couple of alert theorists—A. G. Doroshkevich and Igor Novikov—very nearly put two and two together. "They knew about Alpher and Herman's prediction of the fireball radiation," says Wilkinson, "and they also had identified the antenna at Bell Labs as the one antenna in the world that was capable of verifying it."

In 1964, Doroshkevich and Novikov, like their counterparts at Princeton, were pouring over Ed Ohm's papers in the *Bell System Technical Journal* (it seems Russians read the American scientific literature more thoroughly than the Americans!). And they had focused their attention on Ohm's 1961 paper—which contained the first reference to the mysterious radio hiss.

But, having put nearly all the jigsaw pieces together, Doroshkevich and Novikov made a heartbreaking mistake just as they were about to complete the puzzle: they misread Ohm's paper.

Ohm stated that he had measured the "sky" temperature to be a little over three degrees. By this he meant that, when he pointed the 20-foot antenna at the sky and accounted for every source of unwanted radio waves, he was still left with an unexplained residue of three degrees. But the Russian astronomers thought that in calculating his sky temperature Ohm had not removed the temperature of the atmosphere. By a coincidence this was also about three degrees, so when the two astronomers subtracted this from the sky temperature, they ended up with essentially nothing. They therefore concluded that there could be no appreciable background glow in the Universe.

"When we looked at the very same paper, we thought there was a very good chance that fireball radiation was in there," says Wilkinson. "But when Doroshkevich and Novikov looked at it, they came to completely the opposite conclusion!"

The two Russian astronomers relayed their conclusion to their senior colleague, Yakov Borisovich Zel'dovich, one of the world's most eminent cosmologists. He took it as proof that the hot Big Bang was wrong, and in 1965 published a paper in which he said precisely this.

Ironically, another prominent cosmologist, Fred Hoyle, had the previous year concluded that the Universe must certainly have gone through a hot, dense phase at some time in the distant past. It was particularly significant that Hoyle should have come to this conclusion because it was his theory that the elements were cooked inside stars which had been responsible for sinking Gamow's idea that they were made in a hot Big Bang.

But, by the early 1960s, it had become abundantly clear to Hoyle that, though his theory was enormously successful in explaining the origin of the huge majority of the elements, there was far too much helium around. Since the beginning of the Universe there had not been enough time for stars to have made it all.

Hoyle and a colleague, Roger Tayler, concluded that the helium must have been made in either a Big Bang or else a lot of "little bangs" spread all over the Universe. Nature was not simple; the elements had not been made in a single place. They had been cooked inside stars and also during a hot dense phase that the Universe had gone through. An obvious consequence of such a hot dense phase, Hoyle and Tayler realized, would be fireball radiation, and its cooled remnant should still be around today.[1]

So now there were three independent teams in the world that had realized there ought to be a universal microwave background permeating the Universe.

But in 1964, when Hoyle and Tayler submitted for publication a paper on the origin of the Universe's helium, they unaccountably left out the prediction of the cosmic background radiation—despite the fact that they had included it in an early draft. The story of the cosmic background radiation has its missed opportunities on the theoretical side as well as on the observational side.

One person who has thought long and hard about why the discovery of the cosmic background radiation—one of the most important of the twentieth century—had to be made by accident and why there was no earlier systematic search for it is the Nobel-Prize-winning physicist Steven Weinberg. In his excellent popular account of the Big Bang, *The First Three Minutes,* he gives three main reasons why. First, says Weinberg, the prediction of the fireball radiation came out of a theory that was later discredited. By the 1950s, it was clear that most elements could not have been made in the Big Bang, as George Gamow had hoped. Secondly, says Weinberg, the theorists who first predicted the Big Bang radiation

[1]Hoyle and Tayler were well aware of Alpher and Herman's prediction of the afterglow of creation. Their helium argument had simply led them to the same conclusion.

Steven Weinberg, the particle physicist who puzzled long and hard over why it had taken until 1965 to discover the Big Bang radiation when the technology to detect it existed in the 1940s.

were told by radio astronomers that it was quite undetectable. But the most important reason why the Big Bang theory did not lead to a search for the fireball radiation, says Weinberg, was that before 1965 it was extraordinarily difficult for any physicist to really take seriously any theory of the early Universe. It was a failure of imagination again. The temperature and density of matter in the first few minutes of the Universe would be so far removed from everyday experience that it was hard for anyone to really believe that they had actually occurred. As Weinberg says, the mistake of physicists is not to take theories too seriously but not to take them seriously enough.

Interstellar thermometers

There is a further bizarre twist to this tale. It turns out that not only had the Big Bang radiation been predicted long before that fateful phone call from Penzias to Dicke, but it had actually been observed as well. In fact, evidence for the cosmic background radiation had been around for more than twenty-five years. It had even been published in the scientific literature, but nobody had taken any notice.

In 1938, a decade before Alpher and Herman made their prediction of the fireball radiation, Walter Adams, the director of the Mount Wilson Observatory in southern California, turned a telescope on a nearby star in the constellation of Ophiuchus, the serpent holder. He immediately noticed an unusual dip in the star's spectrum. The dip was just what would be expected if some of the light was being absorbed by molecules of a gas called cyanogen.

Now molecules are fragile things. They are easily broken apart by extreme heat, so they tend not to be found close to stars. Adams therefore concluded that the cyanogen molecules he was seeing were in an invisible cloud of interstellar gas suspended in space somewhere between the star and the Earth.

Such clouds of gas are scattered all over the Galaxy—they are the places where stars like the sun are born—so finding one in front of this nearby star was not much of a surprise. But the pattern of the cyanogen absorption was. The only way Adams could make any sense of the pattern was if most of the cyanogen molecules—which are like little atomic dumbbells—were spinning, tumbling end over end as they drifted through space.

But this was impossible. Interstellar space is mind-numbingly cold, within a whisker of absolute zero, the temperature at which all movement slows to a standstill.

Something had to be driving the tiny cyanogen molecules, causing them to rotate. Andrew McKellar, an astronomer at the Dominion Observatory in Canada, calculated what the something had to be. It was radiation at a temperature of about 2.3 degrees above absolute zero and at a wavelength of 2.64 millimeters. But what that radiation was and where it was coming from he had no idea.

Several other stars were found that also revealed cyanogen molecules rotating faster than they should. So the radiation that was buffeting the tiny cyanogen molecules had to be widespread in the Galaxy—if not universal.

No other astronomers considered the anomaly worth losing sleep over so, like so many discoveries made before their time, it

Walter Adams, the director of the Mount Wilson Observatory in California. In 1938, he discovered that cyanogen molecules floating in interstellar space were being constantly buffetted by radiation at a few degrees above absolute zero.

was forgotten. Until 1965, that is. Then several people, including the Russian astronomer Iosef Shklovski, suddenly remembered the work of Adams and McKellar. They pointed out that the tiny molecules that had no right to be spinning in the dead cold of space were spinning because they were being buffeted by the afterglow of the Big Bang. They were made-to-order interstellar thermometers, sitting in space and quietly taking the temperature of the Universe.

At last, the mystery of the cyanogen molecules was solved.

This guy is really sticking his neck out

In June 1965, Jim Peebles gave a public lecture on the fireball radiation at a meeting of the American Physical Society in New York. It was destined to bring home to him for the first time just what a risk the Princeton team had taken in claiming from just one observation that the radiation from the Big Bang had been found.

To illustrate his talk, Peebles had prepared a lantern slide, showing Penzias and Wilson's background measurement at 7.35 centimeters—a single point. Through it, he had drawn the distinctive humped curve of a black body, showing how the Big Bang radiation varied with wavelength. When Peebles projected the slide onto the screen, he was startled by the audience's reaction. "People began to giggle," he says.

Frowning, Peebles looked up at the graph on the screen. And for the first time, he saw it through eyes of other people, realizing with a sudden shock just how ridiculous it must seem. With total confidence, he had drawn a complicated curve through a single data point. He had joined the dots when there was only one dot to join. Even the schoolchildren in Peebles's audience knew that you could draw any curve whatsoever through a single point and they would all be equally correct.

"Immediately people saw the graph, they saw how ridiculous it was," says Peebles. "They were thinking, this guy is really sticking his neck out. I was aware it was pretty speculative to think that this detection was the background radiation," he says. "But what I hadn't reflected on was just how dramatic a prediction we had made and how much room there was to be wrong."

But though he and Wilkinson and the rest had well and truly stuck their necks out, Peebles did not think it had required any particular courage to do so. "If we had been wrong, it would have rolled off us like water off a duck's back," he says.

"You do science by making bold guesses. They're not bold in the sense that you're putting your physical neck on the line. So as long as you don't make too many bold guesses that turn out to be wrong, you're not even compromising your reputation particularly."

The jury was still out on whether Peebles was right or wrong. A decision would be made when more data points had been collected. As the affair of the giggling audience had emphasized, a lot more proof was needed before anyone could be that certain the signal Penzias and Wilson had picked up had really come from the Big Bang.

Relief and disappointment

The Big Bang idea passed its first major test in December 1965 when Wilkinson and Roll finally got their rooftop antenna working, nearly a year after they had begun building it. They successfully measured the temperature of the sky at a wavelength of 3.2 centimeters, and found that it was around 3 degrees above absolute zero, in perfect agreement with what had been found by Penzias and Wilson.

"Getting that result was a great relief for us, because our neck was stuck pretty far out," says Wilkinson. "Our paper in *Astrophysical Journal Letters* got a lot of ridicule. Most people thought our interpretation was pretty wild. I mean, we only had one point!"

But relief was not the only emotion Wilkinson felt when he and Roll finally made that measurement. "I have to admit it was a bit of an anticlimax," he says.

Bell Labs had effectively stolen Princeton's thunder. Back in the spring, Peebles had felt no real disappointment that Roll and Wilkinson had been pipped at the post. But it was different now, seeing that the measurement the Princeton team had made did indeed agree with the Big Bang interpretation, and that Princeton could easily have been first. "That's when I started to think it was a great shame that Dave and Peter were scooped," says Peebles.

The smoothness measurement

So now there were two observations of the background radiation at two wavelengths. It was perhaps not much but it was a start. Both were consistent with the spectrum of the radiation being a black body with a single temperature, precisely what would be expected if the radiation came from the Big Bang.

But there was a second test of whether the background radiation really came from the Big Bang. As well as having a black body spectrum, it should be equally bright in all directions in the sky. "We knew that this stuff had better be smoothly distributed around us," says Peebles.

As soon as the Princeton group had made the second detection of the cosmic background radiation, Dicke thought it would be a good idea to modify the apparatus and try and measure how smooth the radiation was over the sky. Wilkinson was now joined by a new partner.

In the summer of 1965, while he and Peter Roll had been fiddling with their rooftop experiment, Dicke had hired a young physicist named Bruce Partridge. When Partridge joined the "gravity group" from Oxford, his first job was to choose an experiment to work on. Dicke showed him the two that were under way at the time.

The first was an experiment to measure the oblateness of the Sun. This was another of Dicke's pet ideas. If the Sun were found to be oblate—slightly squashed in shape—then, Dicke's own theory of gravity would be just as effective as Einstein's in explaining the orbit of Mercury.

But when Partridge looked at the solar oblateness experiment, he was dismayed. "The whole thing looked horrendously complicated," he says. "There were racks of electronics everywhere." He asked weakly if he could see the second experiment. Dicke took him to see Wilkinson and Roll's microwave background experi-

Bruce Partridge (right) with one of his student assistants on an observing campaign in 1983.

ment, and Partridge immediately breathed a sigh of relief. "It looked so much simpler," he says. "So that's the experiment I chose."

So it was that Bruce Partridge came to be working with Dave Wilkinson on the experiment to measure the smoothness of the microwave background.

In practice, the measurement would involve pointing an antenna at different parts of the sky and comparing the temperatures it registered. There were certain advantages in comparing the background radiation with itself. For a start, you did not have to worry so much about sources of unwanted radio waves since they would often be the same when the antenna was looking in different directions. When one temperature was subtracted from the other to see what was left, these signals would simply cancel out.

Penzias and Wilson had already shown that the background radiation varied in temperature by less than ten per cent as their antenna swung around the sky. "We realized that we could do a hell of a lot better than that," says Partridge.

"We didn't even take our antenna off the roof," says Wilkinson. "Instead of having it pointed straight up, we simply tipped it over to forty-five degrees."

The idea was to point the trumpet-shaped horn at the "celestial equator." This was an imaginary circle on the sky, essentially where the Earth's equator would be if it were extended out to meet the sky. So every twenty-four hours, the rotation of the Earth would swing the antenna through a complete circle. If the background radiation varied around this circle, then the temperature the antenna registered should vary slowly in the course of twenty-four hours.

It sounded straightforward, but in practice there were complications. Other mundane things could also make the temperature recorded by Wilkinson and Partridge's antenna vary every twenty-four hours. For instance, during the day the Sun would heat the horn, causing it to produce more unwanted radio waves than at night. Somehow they would have to distinguish this temperature variation from a real effect in the cosmic background radiation.

In the earlier experiment, Wilkinson and Roll had managed to remove such unwanted effects by occasionally making the antenna look at an artificial source of radio waves kept at constant temperature—a cold load. But this was deemed too cumbersome for the smoothness experiment. Instead, Wilkinson and Partridge periodically slid a vertical metal mirror in front of the horn so that instead of looking at the celestial equator, it looked at a spot on the sky known as the north celestial pole.

Essentially, this is where the Earth's north pole would be if it was extended upwards to meet the sky. The celestial pole never moves; if you had the patience to watch the sky all night, you would see all the stars slowly circle around it.

In their smoothness experiment, Wilkinson and Partridge constantly subtracted the temperature of the sky along the celestial equator from the temperature of the sky at the celestial pole. In both cases, the day-night heating would cause the signal to vary in the same way, so this unwanted effect cancelled out. Also, in both cases, the horn would be looking through the same amount of atmosphere, so the unwanted signal from the atmosphere cancelled out.

But the most important thing about looking in the direction of the celestial pole was that the temperature of background was constant there. In effect, the patch of sky at the north celestial pole was a natural cold load. By constantly looking back and forth between the equator and the north pole, Wilkinson and Partridge were able to map the true temperature variations in the background radiation around the celestial equator.

By 1966, Wilkinson and Partridge had found that the background radiation varied in temperature by less than about a tenth of one per cent around the celestial equator. "We improved Penzias and Wilson's result by nearly a hundred times," says Partridge.

The Big Bang radiation had passed its second major test with flying colors. It was coming equally from all directions, at least as far as the technology of 1966 could tell.

The early Universe becomes respectable

After the discovery of the cosmic background radiation, people began to take the early Universe seriously. Gamow had shown how a knowledge of nuclear physics could help us understand what was going on in the Universe a few minutes after the Big Bang when the temperature was millions or billions of degrees. But what about even earlier times, when the temperature was even higher, the conditions even more extreme? Insight into these remote times would come from a curious marriage between the science of the very small and the very large—between particle physics and cosmology.

Particle physicists want to find out what makes up all of matter. At one time, they thought it was atoms but then they found that atoms are made of other things—protons, neutrons and electrons. Later, to their dismay, they found that even protons and neutrons are made of other things—quarks. Nobody has isolated a quark and nobody is sure whether these particles are really the end of the line. Perhaps the particles of matter are like Russian dolls, and we will constantly find new ones as we probe deeper and deeper beneath the surface of reality.

The early Universe and particle physics are intimately connected because at the high temperatures that existed in the Big Bang particles were flying about so violently that when they struck one another they disintegrated into their constituents. Particle physicists mimic this inside giant particle accelerators, whirling the microscopic components of matter at great speed and slamming them into one another. For a fleeting instant, they can create conditions that have not existed in the Universe since the first split second after the Big Bang.

Gamow knew about nuclear physics—the physics at temperatures of millions and billions of degrees—and applied that knowledge to the early Universe. Today's physicists have learned about the physics at temperatures of trillions of degrees and hotter. However, whereas Gamow probed the era a few minutes after the Big Bang, today's physicists confidently predict the conditions in the first thousandths of a second. In fact, they have gone much further back, although with less confidence.

It may seem audacious for us sitting here on Earth to claim we know what the Universe was like at such a remote time. After all, the Big Bang theory is based only on three pieces of observational evidence—the expansion of the Universe, the existence of the fireball radiation and the abundance of helium. But we can say so much because the early Universe was so simple. It gets hotter and hotter in a predictable way the further back we probe, but at any time we only have to know the temperature and we have completely described the entire Universe. It only remains to put in the physics of particles that would have existed at that temperature and we know everything.

The problem of course is that sooner or later our knowledge of particle physics gets shaky. We simply cannot achieve comparable temperatures on Earth to test it. We are in unknown country. But even here there is now a guide. For the marriage between particle physics and cosmology has shown how they are independent, how many of the features of the Universe must have been determined by the physics of the very small in the earliest moments after the Big Bang. Whatever physics we use it cannot have consequences in the greater Universe that conflict with what astronomers observe all around us.

This is the legacy of Gamow. For we now see that the ultimate questions of where the Universe came from can be answered only by particle physics.

The ultimate seal of respectability

George Gamow died in 1968, so he did not live to see his ideas vindicated. They received the ultimate seal of approval in 1978 when Arno Penzias and Robert Wilson were awarded the Nobel Prize for their discovery of the cosmic background radiation.

Wilson got his first hint about the Prize in early 1978. "Some guy published a prediction of future Nobel Prizes—I think it was in the magazine *Omni*—and he listed us," says Wilson. "But he'd been wrong on a bunch of things so Arno and I didn't take it seriously." In summer 1978, there was another hint, this time from Gerry Wrixon, an Irishman who had worked for a while at Bell Labs before returning to Europe. While visiting Sweden, Wrixon had been buttonholed by one of the country's leading radio astronomers. "Gerry got asked some very detailed questions about Arno and me and our relationship," says Wilson. "Who did what—that sort of thing."

Later, a Swiss colleague of Wilson's named Martin Schneider dropped an even more blatant hint. Schneider was overdue in handing Wilson a progress report on an experiment, so when the two ran into each other in a corridor at Bell Labs, Wilson mentioned the report, asking whether he could have it on his desk the next day. To Wilson's amazement, Schneider said, no. "You won't want it tomorrow," he said gleefully, "because they're going to announce your Nobel Prize!"

"I must admit I didn't take that too seriously," says Wilson. But the next day he was woken by the phone jangling at seven in the morning. It was another of Wilson's colleagues at Bell Labs. He had

Arno Penzias and Robert Wilson congratulate each other after receiving the 1978 Nobel Prize for Physics for their discovery of the "afterglow of creation."

heard a news item on WCBS and wanted to know was it true what people were saying, that he and Arno Penzias had won the Nobel Prize?

Wilson could not say for sure. But finally he received a telegram, saying that the Swedish Royal Academy of Sciences had awarded the 1978 Nobel Prize for physics to Penzias and Wilson for discovering the 3-degree cosmic background radiation. "It wasn't a complete surprise after all the hints," says Wilson.

"I still don't know where Schneider got his information," says Wilson, "but he's the sort who would dig around and investigate things. It was nice there was no hint until the last year. I think it was lucky not to have people saying year after year you're going to get the Nobel Prize."

The Nobel committee had decided to award its Prize to the discoverers of the microwave background rather than to those who had predicted its existence. In this way, they neatly avoided the sticky problem of deciding who in fact deserved the credit.

"I was deeply disappointed that Bob Dicke didn't get part of the Prize," says Peebles. "I think a good solution would have been Penzias, Wilson and Dicke." Gamow was dead, and one of the rules is that Nobel Prizes are never given posthumously.

"I suppose that with all these awards the Nobel people have to make some sort of semi-arbitrary decision. And that's what they did in the case of the microwave background."

Another consideration of the committee may have been that theories are more quavery than experimental results. Certainly, the two physicists who won the Prize for discovering high-temperature superconductors got it within a couple of years, whereas for relativity, one of the towering achievements of twentieth-century science, Einstein never got the Prize.

Wilkinson is absolutely clear on why Penzias and Wilson got the Prize. "They discovered something fundamental and important about the Universe," he says. "Also, they were first-rate experimenters."

The pair underlined this in the late 1960s when they made another major astronomical discovery. "They discovered large amounts of the molecule carbon monoxide floating out in space," says Wilkinson. After molecular hydrogen, carbon monoxide turned out to be the most common molecule in the Universe.

8

THE TOUGHEST MEASUREMENT IN SCIENCE

Twenty-five years of ill-fated experiments

In the spring of 1967, Dave Wilkinson and Bruce Partridge turned their rent-a-truck through the gates of a US army base in Yuma, Arizona. Back in Princeton their experiment to measure the smoothness of the Big Bang radiation had been hampered by clouds of water vapor high in the air above New Jersey. After redesigning it, they had headed southwest to the place they had determined was the sunniest spot in the United States.

"The US army loaned us an area of desert surrounded by a high fence," says Partridge. "It even gave us active ranks so we could use the officers' club during our stay. Dave was a captain and I was a lieutenant."

The site was perfect for the Princeton team's purpose. It was sunny and dry and the high fence stopped any animals that might be around from trampling over the equipment. The site had only one drawback. "It was where the army put their nerve gas shells to see if they would leak in desert conditions," says Partridge.

Under an iron-blue sky, surrounded by racks of nerve gas shells gently roasting in the Sun, Wilkinson and Partridge went to work setting up their equipment. Neither of them gave too much thought to the danger. "If certain symptoms appeared we were told to slap on our masks and get out of the area as fast as possible," says Partridge.

The electronics for the experiment went in a garden hut they had bought from Sears-Roebuck. The microwave horn was stuck out in the desert, pointing down at the ground rather than upwards

Dave Wilkinson and Bruce Partridge's ill-fated smoothness experiment in Yuma, Arizona, "the driest place in America." The dark object pointing downwards to the right of the Sears-Roebuck shed is the horn antenna for collecting microwaves from the Big Bang. Surrounding it are screens to prevent microwaves from the ground from getting into the horn.

at the sky. "If we'd left the horn pointing up, it would have soon filled up with dead bugs and water from condensation," says Partridge. A metal mirror placed underneath the horn ensured that radio waves from the sky were reflected into the horn's flared opening.

For several weeks, Partridge and Wilkinson worked among the nerve gas shells, riding about the desert on a moped they had bought with some of their research money. "The moped was a lot cheaper than a rent-a-car," says Partridge. "It saved the taxpayers a lot of money." The experiment was designed to be automated so when they had finished setting things up they left it to chug away on its own in the desert.

Things did not go right. "The Yuma experiment ran for a year," says Partridge. "But it was a complete failure."

It turned out that when the mirror switched the horn from looking at one part of the sky to another, the horn saw different temperatures all right. But the temperature difference was not in the cosmic background radiation. It was more local than that. Different parts of the sky contained different amounts of water vapor and the effect of this unevenness completely overwhelmed any variation there might have been in the radiation from the beginning of time.

"When Dave and I had sat down with weather records back in Princeton, we'd figured out that Yuma, Arizona, was the sunniest place in America," says Partridge. "We were absolutely right—it was sunny. But there was still plenty of water vapor hanging in the air. It just didn't show up as visible clouds."

The failure of the Yuma experiment highlighted just how difficult it was to do cosmic background experiments beneath the thick blanket of the atmosphere. "Our experiments to probe the Big Bang would continually be defeated by the cussedness of nature," says Partridge.

It would become a depressing pattern over the next two decades.

Taking a peek above the atmosphere

Moisture hanging invisibly in the air proved to be the bane of cosmic background experiments. It was particularly troublesome when observations were made at short wavelengths of a few millimeters or less. Here water vapor glowed so brilliantly that it overwhelmed the precious cosmic background radiation. Unfortunately, is was precisely at these short wavelengths that astronomers most wanted to observe the afterglow of creation.

The reason was that a black body at a temperature of about three degrees above absolute zero had a peak in its spectrum at a wavelength of around a millimeter. To prove once and for all that the cosmic background radiation really had come straight from the Big Bang astronomers would have to prove that its spectrum was black-body-shaped. In practice this meant looking for the peak and showing that beyond the peak the spectrum fell sharply away.

At a wavelength of about a millimeter, water vapor and other molecules in the atmosphere were glowing fiercely. Beyond the peak, at "sub-millimeter" wavelengths, the situation was even worse. Not only would water vapor be shining brightly but the cosmic background radiation itself would be getting rapidly fainter. The tiny signal from the background would be completely swamped.

Observing the cosmic background at the peak of the spectrum and beyond was a formidable problem. The only solution was to take an instrument to high altitude and get above as much of the obscuring atmosphere as possible. As anyone who climbs up a mountain knows, it gets colder as you get higher. If you go high enough it gets so cold that the water vapor in the air turns to ice and simply drops out as snow.[1]

In their quest to steal a march on the atmosphere, researchers would go from deserts to mountain tops to high-flying balloons, spy planes and rockets. And finally one day they would even go into space.

The peak and beyond

After the Yuma debacle, Wilkinson and Partridge turned their attention back to measuring the spectrum of the cosmic background radiation and to confirming that it was indeed a black body. They had learned a painful lesson in the desert of southwest Arizona. They would not make the same mistake again.

For the site of their new experiment, they selected the summit of White Mountain in northern California. At 12,500 feet, it was the highest and driest spot in the United States.

Wilkinson and Partridge were not the only ones to notice this. "When we arrived at White Mountain and drove onto the site we

[1]Actually, there is one place on Earth with virtually no water vapor in the air, even at ground level: the continent of Antarctica. The air down there is simply too cold for water vapor to exist.

discovered a suspicious-looking device with a microwave horn," says Partridge. "Bernie Burke and some people from MIT were on the summit doing precisely the same thing as us!" It went to show what a boom industry measuring the cosmic background was in the early days—before it began to get hard.

Wilkinson and Partridge were now working with a colleague, Bob Stokes. They had brought three microwave horns with them, each operating at a different wavelength. The plan was to kill three cosmological birds with one stone and pin down a trio of points on the spectrum of the cosmic background.

In the wake of Penzias and Wilson's discovery, every astronomer in the world who had access to a suitable radio telescope had attempted to measure the cosmic background. By the middle of 1966, the temperature had been shown to be close to 3 degrees all the way from 21 centimeters to 2.6 millimeters, a span of almost a hundred times in wavelength.

But all the ground-based measurements had been made on only one side of the humped spectrum—the side at relatively long radio wavelengths. With their three microwave horns, the Princeton team intended to probe the spectrum at the peak and at the shorter wavelengths beyond. They worked like beavers on their experiment for a month and a half. And this time they were rewarded with success. "We found the first tentative evidence of a turndown after the peak," says Partridge.

But it was the end of the road for conventional microwave technology. It was impossible to build microwave receivers at wavelengths as short as a millimeter. Water vapor was a known problem at short wavelengths. But the main reason radio astronomers had first filled in the long-wavelength side of the humped spectrum was because they were able to make use of tried-and-tested microwave receivers. The long-wavelength side of the spectrum was the easy side.

"Now the easy cream had been skimmed," says Partridge. "What was left was hard."

After the early frenzy of activity, there would be a fallow period. "At Princeton, at least, people went off to do other things," says Partridge. He himself turned to more conventional radio and optical astronomy.

In the early 1970s, very little was being done. To go any further would need entirely new technology: the technology to detect radiation at millimeter and sub-millimeter wavelengths.

Bitten by the background bug

One person who soldiered on in the field even during the hard times was Dave Wilkinson. That day in 1965, when Bob Dicke had burst into his lab and announced that the Universe might be filled with the afterglow of creation, had been a fateful day. Wilkinson had been bitten by the background bug and he was infected for life.

He was not alone. The people who do cosmic background experiments are a dedicated band. They tend to stay in the field for the rest of their careers. Even Partridge, who went off to do other things, would come back to the cosmic background, again and again.

Partridge knows exactly why he is so fascinated by the cosmic background radiation. "For me the answer is quite clear," he says. "Its simplicity."

"The experiments to measure the radiation are simple to understand and simple to describe. The radiation itself is simple—it's a black body and it has the same temperature in all directions in the sky. Once you've said those two things about it, you've said everything there is to say."

"The simplicity of the cosmic background radiation is telling us something marvelous—that the early Universe was a remarkably uncomplicated place," says Partridge.

"It's the only way to look back to the beginning of the Universe," says Bob Dicke.

Wilkinson agrees. But the cosmology is only part of the reason he is fascinated. The main reason is that he loves the challenge of designing and building experiments to outwit nature. "The experiments are the sort I like," he says. "They're tough but important. You have to think hard about the unwanted effects and there's a novelty and cleverness in experiments."

It goes back to his childhood. "I'm a tinkerer," says Wilkinson. "I got it from my dad. He only graduated from high school but got interested in electronics. He had a workshop in our basement. When I was a kid, I was always tinkering with cars and electronics."

But the appeal for Wilkinson does not end here. There is another thing that has always appealed to him. "You can carry out an important experiment with just you and a graduate student," he says. "You have complete control over an experiment. It's small-scale, manageable science."

Nowadays, when so much science is big science, and carried out by international teams of hundreds of scientists, it is easy to see why background work appeals so much.

Balloons and rockets and planes

During the early 1970s Wilkinson took advantage of a new technology that was being developed: that of high-altitude balloons. "It was the failure of the Yuma experiment that led Dave to first think about using balloons to get above most of the water vapor in the atmosphere," says Partridge.

Balloons could take an instrument package to an enormous height—perhaps three of four times the height of Everest. The air at such a height would be so thin that the instruments would practically be in space. For ten hours or so, before the strong winds at such high altitudes blew the balloon over the sea or simply out of radio range, the instruments could get an almost unobstructed peek at the Universe.

Hoisting experiments aloft on balloons meant building cosmic background experiments that were a lot more complicated than before. Everything had to be done remotely. Even the simple operation of sticking a cold load in front of an antenna—so easy on the ground—was difficult and plagued by problems when it had to be automated.

"Balloon experiments are not like table-top experiments," says Wilkinson. "If you find an error in one of those, you can modify the experiment and do it again. You can't do that with one balloon flight a year."

Unexpected things were likely to happen in the extreme environment twenty or thirty miles up in the air. For a start, it would be dreadfully cold. Ice could freeze up the equipment. All sorts of

David Wilkinson (center) and his team at the National Scientific Balloon Facility in Palestine, Texas, preparing to launch an early balloon experiment to measure the smoothness of the cosmic background radiation.

things could happen that would be easy to put right on the ground but that could wreck an experiment flying on its own beneath a balloon on the edge of space.

On balloon campaigns, Wilkinson has had a secret weapon—his father. Because he was retired and lived in Texas where the balloons were launched, he often turned up to lend a hand. Wilkinson, usually working with just one graduate student, was always grateful. It took lots of work to ready a payload for launch.

Balloons were only one of the ways people found to get above the atmosphere. Some used high-flying airplanes. Others used sounding rockets. These pencil-thin launchers were used by meteorologists for studying the upper atmosphere. They went straight up to a height of a few hundred miles then came right back down again when their fuel ran out. But in the few minutes they were in space the instruments they carried got a clear view of the

Universe. The drawback was that everything had to work perfectly during those few minutes, or years of hard work went down the drain.

The new astronomy

What kept the field alive during the 1970s were the efforts of researchers designing new technology to bring to bear on the microwave background. At the beginning of the decade, there was a major breakthrough in detector technology. New detectors came in that were called "bolometers." These responded to the warmth of incoming radiation. Essentially, a very small amount of heat changed their resistance to an electric current by a large amount and this was something experimenters could easily measure.

Bolometers were much better than radio receivers at detecting faint radiation at short wavelengths of a few millimeters. And they worked at even shorter wavelengths where it was impossible to build radio receivers. To achieve the best results, though, bolometers had to be cooled to within a whisker of absolute zero.

Into the cosmic background field came groups of scientists specializing in using the new detectors. Rai Weiss and his colleagues at Boston's Massachusetts Institute of Technology were among the first to use bolometers in the early 1970s. In Britain, a group led by John Beckman and Peter Clegg at Queen Mary College in London made a foray into the field. Another group got started at the University of California at Berkeley. It included Paul Richards, John Mather and George Smoot. Both American groups were destined to have an important impact on the field.

But the first observations of the cosmic background radiation made with bolometers did not show the drop-off with wavelength expected for a black body. "The experiments were extremely difficult to do," says Wilkinson.

Rather than responding to one wavelength like radio receivers, bolometers responded to all wavelengths at once. This meant that to make a measurement at any one wavelength, it was necessary to put a "filter" in front of a detector. The filter was transparent to the wavelength of interest, but absorbed all other wavelengths.

A familiar example of a filter is a sheet of red cellophane. This allows through red light while stopping, or "filtering out," all other colors. Similarly, a blue filter is transparent only to blue light.

For observing the cosmic background radiation, scientists like Rai Weiss used a bolometer in conjunction with several "filters." But though it was now possible to make measurements that were quite impossible with microwave receivers, instruments using bolometers were not without their problems.

For a start, filters absorbed radiation, and anything that absorbs must also emit (otherwise it would simply get hotter and hotter until it was white-hot). So filters were yet another source of unwanted radiation which astronomers had to learn to contend with.

Another problem with filters was that they threw away a lot of the precious cosmic background radiation, allowing through only light at one particular wavelength. This was very wasteful. But in the late 1970s there was another major development in the field. The second generation of experiments at short wavelengths used an instrument known as the Michelson interferometer, which allowed all wavelengths to be detected at once.

The Michelson interferometer had been invented back in the 1880s by the American physicist Albert Michelson.[2] Essentially all such an instrument does is split the light into two equal parts, then recombine it. This may seem a pretty pointless thing to do but, after the light is split and before it is recombined, the two halves are made to travel different distances—usually by bouncing them off two separate mirrors. This path difference can be changed continuously by gradually moving one of the mirrors.

Since the light entering a Michelson interferometer consists of waves of a multitude of different lengths mixed in together, when the light is recombined, each different wavelength combines with its other half. For a particular wavelength, if the peaks of the two halves coincide despite having travelled different distances, then the waves reinforce each other when they recombine. But if the peaks of one half coincide with the troughs of the other, then the waves cancel each other out.

---

[2]In 1907, Michelson became the first American to win a Nobel Prize.

The wavelengths that reinforce and cancel will change as the path difference is changed. So when the recombined light falls on a bolometer detector, the brightness that registers will vary. The way in which the brightness varies with path difference is known as an "interferogram." In theory, the interferogram contains all the information needed to determine the brightness of the light at a large number of wavelengths simultaneously—in other words, to determine the spectrum of the light. In practice, it takes a bit of mathematical manipulation to extract the spectrum.

The details of how you do this do not concern us here. The important thing is that a Michelson interferometer can measure all wavelengths at once, and so waste none. And it has another feature, which makes it particularly suited to measuring the spectrum of the cosmic background radiation: it can compare the sky and the cold load all the time. There is no need to keep switching back and forth between the sky and the cold load, and wasting half the precious light from the sky.

By the late 1970s, Michelson interferometers equipped with bolometers and cold loads, both cooled by liquid helium, represented the state of the art in measuring the spectrum of the cosmic background radiation. But in the 1980s experimenters took a final step. Instead of using a small amount of liquid helium, they used a very large amount. In fact, they immersed their entire instruments, including their antennas, in large vacuum flasks full of the liquid. This cooled them to within a few degrees of absolute zero, dramatically reducing the unwanted radiation from the instruments themselves.

The hardest accounting job

But despite the advances, measuring the cosmic background radiation remained tough. It was, after all, the lowest temperature in the Universe. Everything else was hotter—the ground, the sky, even the instrument making the measurement. So whenever they measured the background all those other things would be in there too. It was like trying to observe a faint star while standing on a searchlight.

Basically, cosmic background experiments came down to good accounting. Measure a temperature. Then think of all the possible confusing effects and estimate how big they are. Better still, go out and measure how big they are. The cosmic background is the residue, what is left over after everything else has been accounted for.

In theory it was straightforward. That was why Partridge said the experiments were simple to understand and describe. But in practice it was a lot harder. "The problem is, have you thought of everything?" says Wilkinson. "The nature of the field is funny," says Wilkinson. "You see someone's result and you have to question everything."

Always something was overlooked

It was almost impossible to think of every spurious source of radiation. Always something was overlooked. Some astronomers accidentally measured the temperature of the plastic window through which their instruments looked at the sky. Others measured the temperature of the balloon that was hoisting their experimental package aloft.

Partridge remembers an experiment done with Norwegian colleagues at a site in northern Norway. It was 250 miles above the Arctic Circle. "We reasoned that if you went high in the Arctic, it was just like going into space," says Partridge. "It's cold, dark and, when there's no Sun, the temperature is stable."

They reasoned wrong. "It was a total failure," says Partridge. "We were killed off by the atmosphere." They had not reckoned on a waterfall of cold air draining off the Russian plateau and down over the Norwegian coastal plateau. "We ended up observing under all this turbulence."

"It turns out the South Pole, not the North, is the place to go," says Partridge. "It's high and dry and the air behaves because the atmosphere there is very stable."

On another occasion, Partridge and his colleagues were flying a balloon experiment to measure the smoothness of the background and ended up measuring the temperature of some cables attached to the balloon.

The experiment had a horn spinning beneath a balloon. Unknown to Partridge's team, the launch cables dropped down—the umbilicus of the balloon. So six rubber-coated wires dangled in front of the horn.

"We'd worried ourselves sick about the four metal cables that suspended the instrument package from the balloon," says Partridge (in the end, they had carefully shielded them). "But nobody told us this was going to happen. Every time the horn swung round it got a huge jolt of radiation emitted by the cables," says Partridge. "On our chart recorder, we got a gigantic trace followed by a long slow decline. The instrument had just about recovered when the horn swung around and saw the cable again!"

But despite all the problems the experimenters were unperturbable. The cosmic background radiation, after all, was our only window on the beginning of the Universe. By the late 1970s, all the hard work was beginning to pay off . . .

9

BUMPS BUT NO WIGGLES

The cosmic background throws up some puzzles

In 1979, Dave Woody and Paul Richards of the University of California at Berkeley used a 120-meter-diameter balloon to hoist a bolometer experiment 27 miles up into the air. Their instrument, dangling 700 yards below the balloon, looked at the cosmic background radiation for three hours before the strong winds at such a high altitude blew it out over the Gulf of Mexico and the balloon had to be recovered. But in those three hours, Woody and Richards's instrument made the best measurement of the spectrum of the Big Bang radiation so far achieved.

The experiment incorporated all the most up-to-date features. Light was collected from the sky by a trumpet-shaped horn, specially designed to keep out stray radiation from nearby warm objects. This was supplemented by an "Earthshine shield" to keep out radiation from the Earth below. The horn funnelled the light down into a Michelson interferometer complete with sensitive bolometer detectors cooled by helium-3 to just 0.3 degrees above absolute zero.[1] These compared the temperature of the sky with an artificial black body cooled by liquid helium.

To reduce unwanted radiation from the apparatus itself, Woody and Richards immersed their entire instrument, including the light-collecting horn, in a vacuum flask of liquid helium. It hung 700 yards beneath their balloon in order to minimize the chance of picking up unwanted radiation from the balloon itself.

[1]Helium-3 is a rare type of helium which boils at an even lower temperature than the common variety.

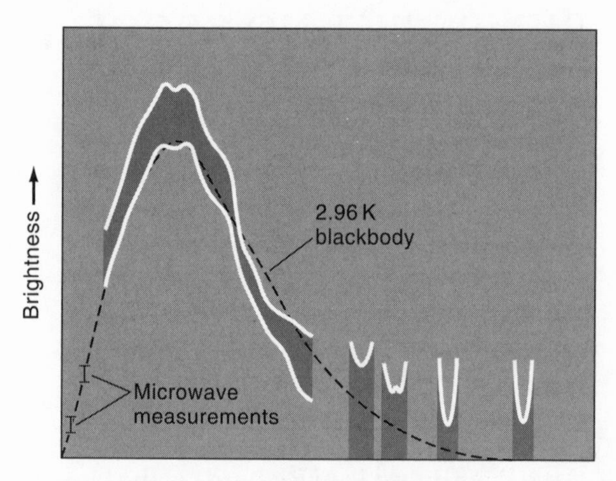

Progress in measuring spectrum of the cosmic background radiation up to 1979. The shaded region between the solid curves shows the measurements of Dave Woody and Paul Richards of the University of California at Berkeley. The bracketed points are some of the early radio measurements. The dotted curve is the spectrum of a theoretical black body at a temperature of 2.96 degrees above absolute zero.

The two astronomers found that the fireball radiation was approximated by a black body at a temperature of 2.96 degrees above absolute zero. "That was the first spectrum I really believed," says Dave Wilkinson.

But there was an important difference between the spectrum Woody and Richards measured and the theoretical 2.96 degree curve. Although at long wavelengths Woody and Richards's observations hugged the 2.96 degree black body curve very closely, at the shorter wavelengths—beyond the peak in the spectrum—the agreement was not nearly so good. There was too much radiation at short wavelengths. The spectrum of the cosmic background radiation appeared to have a bump in it.

"It was a pattern we were to see repeated several times in the field," says Bruce Partridge. "Whenever people measured the spectrum of the Big Bang radiation, the measurements at short wavelengths—less than a millimeter—always showed a puzzling excess."

While some persevered with balloon-borne experiments, others tried firing instruments high into the atmosphere on the tip of sounding rockets, pencil-thin launchers more often used by meteorologists to study atmospheric phenomena such as the aurora. One person who specialized in this sort of experiment was Herb Gush at the University of British Columbia in Vancouver. During the 1970s, he fired a number of such experiments up to a height of a few hundred miles, and he, too, measured an excess of Big Bang radiation at short wavelengths. But Gush's experiments were plagued by problems—for instance, glowing gases from his rocket exhaust had a habit of obscuring his instruments' view of the Universe, making his measurements questionable.

But in 1988, Paul Richards at Berkeley joined forces with a Japanese team led by Toshio Matsumoto at Nagoya University. Together they launched a rocket experiment, which obtained a spectrum of the cosmic background that most people in the field believed. Since it, too, had a bump with a wavelength of about a millimeter, there was intense interest from the theorists. They came up with a flurry of possible explanations.

Because the hot Big Bang naturally produced fireball radiation with the spectrum of a perfect black body, the bump in the spectrum had to mean that some process since the Big Bang had injected an immense amount of energy into the Universe. There were all sorts of possibilities.

For instance, there might be large amounts of warm dust suspended in galaxies or else hanging between the galaxies, and the glow of this dust might be responsible for the bump in the cosmic background spectrum. But the problem here was to find a way of heating up the dust so that it glowed brightly at around a millimeter in wavelength. The dust had to be pretty much everywhere in the Universe since astronomers knew that the background radiation came equally from all directions. Clearly, a prodigious amount of energy would have been needed to warm it up.

The theorists thought of many possibilities. For instance, a large number of stars might have formed shortly after the Big Bang.

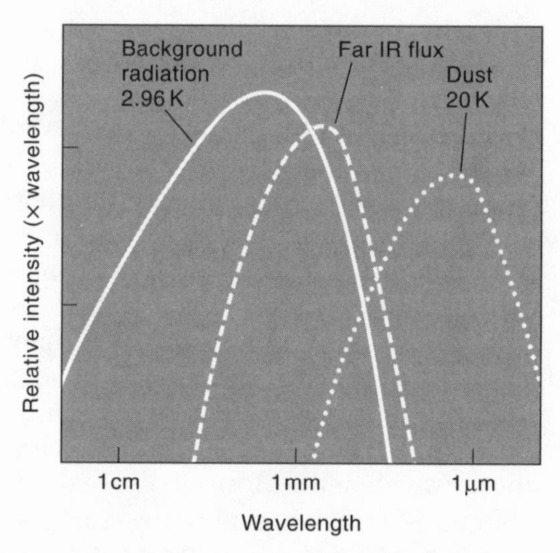

An anomalous "bump" in the cosmic background spectrum at a wavelength of about a millimeter was found by Toshio Matsumoto and Paul Richards in 1988. Theorists tore their hair out trying to explain it.

They could have raced through their life cycles and exploded, giving out an enormous quantity of energy. Another possibility was that the early Universe contained "exotic" microscopic particles, as yet unknown to science, and that these had decayed since the Big Bang, releasing a lot of energy.

But all these schemes failed ultimately. The energy required to heat up such a large quantity of dust in the Universe was simply too enormous. The theorists could imagine no plausible physical process that would work.

But some considered an alternative. They wondered whether the bump in the spectrum could be explained if some process had simply redistributed the energy in the fireball radiation, sapping it from long wavelengths and redepositing it at a wavelength of around a millimeter to create the observed bump in the spectrum.

In this scheme, the photons of the background radiation had not after all flown unhindered across space for 15 billion years but had instead passed through clouds of extremely hot gas floating between the galaxies. Such hot gas would be permeated by fast-moving electrons, stripped from the gas atoms. If the photons of the cosmic background collided with these electrons, they would rob them of energy, boosting their own energy and shortening their wavelength. The net effect of countless collisions would be to take energy from the background radiation at longer wavelengths and deposit it at wavelengths shorter than the peak. It would create precisely the bump seen in the Berkeley-Nagoya experiment.

But there were problems with this idea, too. For a start, no one knew whether such a hot gas existed throughout the Universe. But the biggest problem with the hot gas scheme was that it ran into the same difficulties as the dust idea. Some way would have to be found to heat the hot gas between the galaxies, and no one could think of where such a prodigious amount of energy could come from. The theorists were flummoxed. The experiments to measure the cosmic background radiation had thrown up an apparently insoluble problem.

The background radiation reveals the motion of the Earth

But it is not only the experiments to measure the spectrum of the cosmic background radiation that were puzzling theorists. The experiments to measure the smoothness of the radiation were also beginning to baffle them.

However, one discovery, made in the late 1970s, had been entirely expected, and that was that the radiation was slightly hotter in one direction in the sky than in the opposite direction because of the Earth's motion through the background radiation.

"The discovery was made incrementally," says Wilkinson. "A series of ground-based experiments, including ours, saw something marginally, then finally the Berkeley group saw it from a high-flying U2 plane."

What Rich Muller, Phil Lubin and George Smoot of Berkeley found in 1977 was that one half of the sky was about 0.1 per cent hotter in the direction of the constellation of Leo than it was in the opposite direction. This amounted to a difference of just three thousandths of a degree. No wonder it had taken more than a decade to find it!

The temperature difference could be explained if our Sun was flying through the cosmic background radiation at a speed of approximately 230 miles per second in the direction of the constellation of Leo. The radiation in the direction we were moving would naturally be blue-shifted by the Doppler effect, boosting it in energy and temperature. The radiation behind, on the other hand, would be red-shifted and reduced in temperature.

"Jim Peebles had told me there would be such an effect even before Penzias and Wilson discovered the Big Bang radiation," says Dave Wilkinson. Peebles had realized that the cosmic background radiation was a sort of universal "frame of reference" against which the speed of every object in the Universe could be measured. Peebles had even predicted roughly how big the effect should be since he knew how fast a typical galaxy like the Milky Way was moving.

Astronomers referred to this observation that one half of the sky was hotter than the other as the "dipole effect." "Seeing the dipole was a major relief," says Wilkinson. "If it hadn't been there, it would have been a major embarrassment to everyone."

The birth of galaxies

But what was destined to become an embarrassment was the fact that apart from the "dipole" variation, the cosmic background radiation seemed utterly smooth across the sky. "We knew the radiation had to be smooth," says Jim Peebles, "but we knew it could not be dead smooth because today's Universe is lumpy."

At some point the smoothly distributed matter in the early Universe had to start clumping to form galaxies and clusters of galaxies, and this should make itself visible as an unevenness in the cosmic background radiation.

Back in 1965, when Bob Dicke had first introduced to him the idea of a hot Big Bang, Peebles had realized that the fireball radiation was linked with the origin of the galaxies such as the Milky

Way. "It was pretty evident that the radiation would have an important effect on how galaxies form," says Peebles.

The fireball radiation completely dominated the Universe during the first 300,000 years after the Big Bang. For every particle of matter, there were between 100 million and a billion photons, a ratio that has remained constant in the Universe to this day. But though today's background photons have been cooled and diluted by the expansion of the Universe, in the early Universe the photons were immensely hot and packed closely together. This meant that in any cubic centimeter of the early Universe the total energy of the photons was enormously greater than the energy of the particles of matter. Matter was only a minor contaminant. In the early Universe, it was radiation that called the shots.

The implication of all this for galaxy formation is that the process could not begin earlier than 300,000 years after the Big Bang. Any particles that came together would simply be blasted apart by photons of the fireball radiation. But at 300,000 years, atoms formed and mopped up all the free electrons through which the photons of the fireball radiation were influencing matter. The universe became transparent to photons, and from that moment on matter and radiation went their separate ways.

Coincidentally, this was also roughly the time that the energy density of radiation in the Universe dropped below that of matter. This happened because the energy of photons was diluted as their wavelength was stretched by the expansion of the Universe. But the energy density of particles of matter cannot be diluted indefinitely because each particle has a floor—a so-called rest energy[2]—below which it cannot go.

So about 300,000 years after the Big Bang, the Universe became dominated by matter. Freed from the tyranny of radiation, matter could begin to clump under the force of gravity. Gravity, not the pressure of radiation, was now the dominant force in the Universe.

Because the photons of the cosmic background radiation last interacted with matter around this time, they ought to reveal how

[2]The rest energy of a particle of mass m is given by Einstein's famous equation $E = mc^2$, where c is the speed of light.

matter was spread throughout the Universe at the time. As early as 1968, the theorist Joseph Silk had pointed out that mapping the temperature of the cosmic background radiation would allow us to "see" clumps of matter 300,000 years after the Big Bang, just as the process of galaxy formation was beginning.

"The lumps would be of exceedingly great interest—that was obvious right away at the beginning," says Peebles.

Where the matter of the early Universe was ever so slightly denser than its surroundings, photons would have to climb out of the slightly stronger gravity. They would lose energy, becoming red-shifted. The gravitational effect, predicted by Einstein in 1915, would create a "coldspot," a region of the sky where the cosmic background radiation was marginally cooler than elsewhere. Similarly, "hotspots" would mark regions of the early Universe that were ever so slightly less dense than average. In effect, the radiation would carry with it an imprint of the Universe as it was soon after the Big Bang itself.

The Universe is made of Swiss cheese

"It took a long time for experiments to measure the smoothness of the Big Bang radiation to catch on," says Wilkinson. "For a good ten years, nobody was doing anything—except Bruce Partridge and me."

But when other astronomers did get involved, they searched in vain for any slight departure from complete smoothness. There were false alarms. An Italian team led by Francesco Melchiorri announced finding hotspots in the cosmic background radiation. So, too, did a team led by Rod Davis of Jodrell Bank in England. They were running an experiment on top of a mountain in the Canary Islands. Both teams had to retract their findings after admitting they had made a mistake.

Even Dave Wilkinson was not immune from such errors. "We reported seeing something at the same time as the Italians," he says. "But we were fooled by radio emissions from our Galaxy."

By 1989, after more than two decades of painstaking observations, astronomers had still not detected any variation in the temperature of the Big Bang radiation across the sky, apart from that due to the relative motion of the Earth. The uniformity seemed to be indicating that when the radiation was produced, about 300,000 years after the Big Bang, the matter of the Universe was spread out completely evenly. This posed a very awkward question because the distribution of matter in the Universe today is anything but uniform. How then did the galaxies and clusters of galaxies in today's Universe form?

By the late 1980s, this question was beginning to give theorists serious headaches. It was not only that the experimenters were finding that the cosmic background radiation and therefore the matter of the early Universe were spread remarkably smoothly. Simultaneously, astronomers mapping how galaxies are spread through space were finding that the matter of today's Universe is spread out a lot more unevenly than anyone had suspected.

These astronomers were making use of sensitive electronic light detectors known as charge-coupled devices or CCDs. Their introduction during the 1970s had brought about a major revolution in astronomy that had never hit the headlines. CCDs were far superior to the photographic plates that astronomers had traditionally used with their telescopes to probe the Universe. Instead of picking up about one per cent of all photons of light collected by the mirror of a telescope—which was typical of photographic plates—CCDs could pick up nearly a hundred per cent of all photons. Just by swapping photographic plates for CCDs, any telescope could instantly be made about a hundred times more sensitive than it was before. And this meant it was possible to study galaxies that were much fainter and therefore farther away than any seen until now.

Using big telescopes equipped with CCDs, some astronomers mapped how galaxies were spread throughout a large volume of the Universe. What they found was that the Universe was full of complex structures. Galaxies are clustered in great chains and sheets that surround great voids of empty space where there are no galaxies to speak of at all: a structure remarkably similar to Swiss cheese.

The origin of these clusters and voids was one of the greatest problems of cosmology. And it was extremely difficult to square

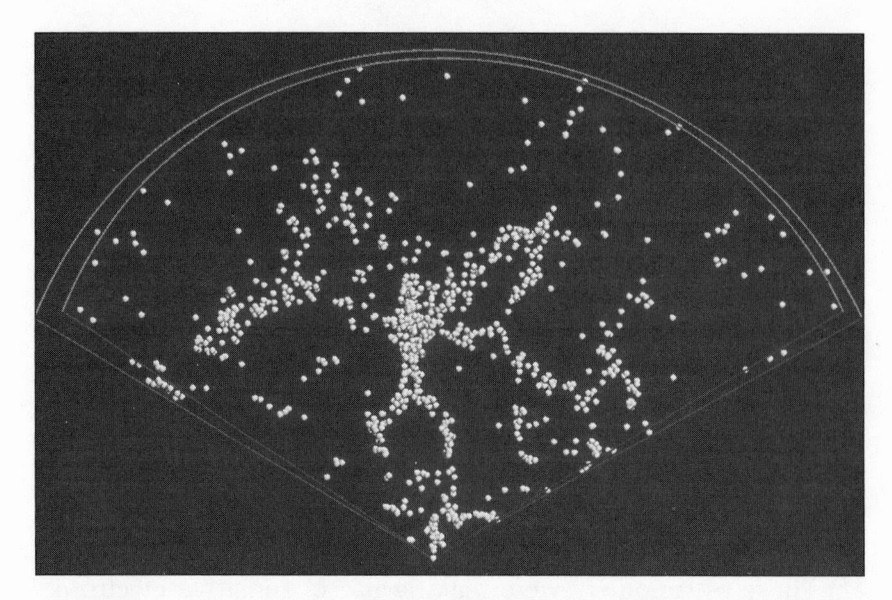

The Universe is made of Swiss cheese. In this slice, each of the 1,000 points is a galaxy like our own Milky Way. The Earth is at the bottom of the slice and at the outer boundary galaxies are receding from us at 7,500 miles per second.

with the evidence of the cosmic microwave background, which was telling us that the early Universe was astonishingly smooth. How could such complexity have come out of such simplicity? The evidence of the cosmic background radiation was saying that by rights our Milky Way should not exist.

Other astronomers probing the depths of space with CCDs discovered objects at greater and greater distances. These were quasars, the phenomenally bright cores of newborn galaxies. No one knows precisely what causes quasars to shine so brightly but because they do they can be spotted at immense distances. By the early 1990s, quasars were being found that were so far away that their light had been travelling to us for most of the history of the Universe. In fact, we were seeing some of them as they were within a billion or so years of the Big Bang.

Again, these observations were extremely difficult to square with the evidence of the cosmic background radiation. How could quasars have condensed out of the cooling fireball within a billion years or so when the fireball radiation was showing no sign of lumpiness?

An eye above the atmosphere

The cosmic background radiation had thrown up two baffling puzzles—it seemed to be too smooth by far and its spectrum had a peculiar bump in it that no theorist could explain. It had taken nearly twenty-five years to reach this point and progress was painfully slow. If the twin puzzles were ever to be solved, it would be necessary to get above the glowing atmosphere. It was only a thin layer, comparable to the thickness of the skin on an apple, yet it was standing between the astronomers and the greatest prize in cosmology. It was clear to everyone that what was needed to solve the puzzles of the fireball radiation was an eye above the atmosphere. What was needed was an experiment in outer space.

PART

II

THE GOLDEN AGE OF COSMOLOGY

10

AN EYE ABOVE THE ATMOSPHERE

NASA decides to nail the problem once and for all

New York's Goddard Space Science Center is a singularly unromantic place. But on a summer day in 1974, this bleak office block in upper Manhattan was the venue for a meeting with more than just a little romance about it. The seven men who came together that day were proposing nothing less than a quest for the Holy Grail of Cosmology.

The catalyst was John Mather, a tall, lean young astronomer barely six months out of graduate school. Before moving east to New York he had cut his cosmological teeth on balloon experiments with Paul Richards at Berkeley. What had prompted him to arrange the meeting at Goddard was Announcement of Opportunity AO6&7, a request by NASA[1] for proposals for new space missions.

"It was obvious to everyone that a satellite would have enormous advantages for probing the microwave background," says Dave Wilkinson, one of the seven astronomers at the meeting. Not only would the instruments on board be able to peek at the Universe from above the weather, but they would be relentless, sitting in orbit for months on end, soaking up the precious photons of background radiation. "A satellite would well and truly nail the problem," says Wilkinson.

[1]The National Aeronautics and Space Administration.

John Mather, the driving force behind the COBE *satellite since 1974.*

Over the next few months, Mather's team, which included Wilkinson and Rai Weiss of the Massachusetts Institute of Technology, hammered out a proposal for a satellite that would carry four separate experiments into the Earth's orbit. The experiments would include one to measure the spectrum of the Big Bang radiation better than anyone had ever dreamed, and another to scan the whole microwave sky, searching for the tiniest departure from absolute smoothness.

"I was only twenty-eight years old at the time we put in the proposal, so I can't say I really took it seriously," says Mather. "But I knew one thing—the idea behind the experiment was a good one."

An idea whose time had come

NASA was deluged with hundreds of space proposals. But Mather's team was in luck. Already, several of NASA's own science committees had identified a space experiment to observe the birth of the Universe as just the sort of project the Agency ought to be carrying out. Such experiments were impossible from the ground and could be done only from space. They were also of fundamental importance to science, dealing with nothing less than the origin of the Universe in the most gigantic explosion of all time.

NASA was also shrewd enough to realize that such a project was likely to capture the imagination of the public. "Everyone wants to know how we got to be here," says Mather. "And that was precisely the question we aimed to answer."

But unknown to Mather and his team, others had also set their sights on the Big Bang. Among the sackloads of space proposals delivered to NASA headquarters were two others for putting a satellite in orbit to probe the faint afterglow of creation.

One proposal came from a team led by Samuel Gulkis of the Jet Propulsion Laboratory in Pasadena, California; the other from Luis Alvarez and his colleagues at the University of California at Berkeley.

Alvarez was a legendary physicist: a Nobel Prize winner who during the war had worked on the development of both the atomic bomb and radar. In a remarkably diverse career he had carried out a search for undiscovered chambers inside Chephren's pyramid, using natural x-rays from space, and started his own company to make variable-focus contact lenses.[2]

That a scientist of the caliber of Alvarez had zeroed in on a measurement of the Big Bang radiation only served to underline how important it was to science.

Both Alvarez and Gulkis wanted to look for tiny departures from smoothness in the microwave background—a single element of Mather's more ambitious proposal. "It was immediately clear we overlapped," says Mather.

NASA, forced to choose between rival cosmic background proposals, came up with a neat solution. It formed a study team with investigators chosen from each of the three proposals.

At this point, Alvarez decided to drop out. The project was now more ambitious than the one he had originally envisaged and, at sixty-five, he had a strong suspicion that he might not last the duration.[3] In his place he nominated a young Berkeley researcher named George Smoot.

Smoot was destined to become the team's most controversial figure. But in the mid-1970s that was all a long time in the future.

---

[2]In 1980, Alvarez would hit newspaper headlines all over the world by claiming to find evidence that a giant impacting asteroid wiped out the dinosaurs 65 million years ago.

[3]Alvarez was right. He died in 1988, a year before the launch of COBE.

The Cosmic Background Explorer

The satellite, by now christened the Cosmic Background Explorer (COBE) and essentially Mather's proposal, would take three experiments high above the troublesome atmosphere. It would protect them from the light and heat from both the Sun and the Earth, supply them with electrical power and transmit their data back down to the ground.

One instrument—the Far Infrared Absolute Spectrophotometer (FIRAS), a direct descendant of experiments Mather had flown on balloons while at Berkeley—would determine the spectrum of the Big Bang radiation a hundred times better than ever before to see whether it really was a black body.

A second instrument—the Diffuse Infrared Background Experiment (DIRBE)—would search for the infrared glow of the first galaxies to form out of the cooling gas of the Big Bang.

The third instrument—the Differential Microwave Radiometer (DMR)—was descended from instruments George Smoot, Phil Lubin and Dave Wilkinson had flown. It would map the brightness of cosmic microwave background with extraordinary sensitivity, looking for the slightest signs of any unevenness.

In 1976, NASA selected the Goddard Space Flight Center as the focal point for further studies. For the next decade and a half, this sprawling facility at Greenbelt on the outskirts of Washington DC would be headquarters for the COBE effort. The people working there would make the greatest day-to-day contribution to the project.

In 1982, after a series of feasibility studies, NASA finally gave the team the go-ahead to start building COBE. The launch was eventually set for 1989.

Selecting an orbit for COBE was no mean feat. The sensitive instruments would have to be kept out of the Sun as much as possible and during the course of a year they would have to see the whole sky.

The Goddard engineers picked a "polar" orbit that would swing the satellite around from pole to pole and always keep it flying along the boundary between night and day. The orbit would gradually drift so the instruments would eventually see all the sky.

The engineers also had to find a way to control the satellite's orientation as it raced through space. If its sensitive instruments ever pointed anywhere near the Sun or the Earth they would be completely blinded. The "attitude control system" that ensured this never happened was such a tour de force of Goddard engineering that the team came to dub it the "fourth experiment."

Shuttle shenanigans

Once it was decided that COBE should be put into a polar orbit, the launcher for the satellite was dictated. NASA had a very reliable expendable launcher called a Delta rocket that was just right for the job. "Everything in the experiment called out for a Delta launch," says Wilkinson.

But NASA had other ideas. By the 1980s, the Agency had plumped firmly for the reusable Space Shuttle as the workhorse of its space effort. "It had put all its eggs in the Shuttle basket," says Charles Bennett, a researcher who joined the COBE team in 1984. Despite objections from the team, NASA insisted on launching COBE on the Space Shuttle. "The Agency simply didn't want competition from throwaway launchers," says Wilkinson.

At the time, the Shuttle was being launched from Cape Canaveral at the tip of Florida, from where it was impossible to put a payload into a polar orbit. But NASA pointed out that it was planning to build a second launch site for the Shuttle at Vandenberg Air Force Base in the desert of southern California. From Vandenberg it *would* be possible to launch a satellite into a polar orbit.

Being scheduled to take off from a launch facility still on the drawing board was a serious worry for the team. But NASA had spoken and it, after all, was footing the $60-million bill for the project. Mather's team knuckled under and began redesigning COBE for launch on the Space Shuttle.

The changes to the satellite were not trivial at all. "When NASA forced us onto the Shuttle it was a major trauma," says Wilkinson. For one thing, the satellite now needed to have a rocket strapped to it to boost it from the Shuttle's cargo bay to its orbit. The Shuttle could haul itself to a height of only 200 miles above the Earth but COBE's polar orbit was 600 miles up.

A Shuttle launch brought other worries, too. For instance, gases in the Shuttle's cargo bay might contaminate the satellite, and unwanted radiation from these could swamp the faint hiss from the cosmic background. But there was a much needed boost to the morale of the team in January 1983 when NASA successfully launched its Infrared Astronomical Satellite. "IRAS was before us in the queue, so you can imagine, we were very keen to see that satellite go," says Mather.

But there was another reason for wanting to see IRAS succeed. In common with COBE, it carried a giant vacuum flask of liquid helium to cool down its instruments so that they would be more sensitive. Liquid helium, which boils at only 4.2 degrees above absolute zero (−269°C), is not the easiest substance to handle at the best of times. "Nobody had used liquid helium in space before," says Mather, "so people were terrified the technology would not work."

The challenge after Challenger

But the IRAS satellite was a great success, relaying back stunning pictures of some of the coldest objects in the Universe—newborn stars and great curtains of gas and dust hanging across space. "The liquid helium technology passed its most severe test," says Mather. "We all breathed a great sigh of relief."

By 1986, COBE was largely built. But on 28 January, the Space Shuttle Challenger exploded into a thousand flaming pieces in the blue Florida sky, killing all seven astronauts on board. As the horrific pictures of the accident were flashed across the television screens of the world, the COBE project seemed to be in ruins. Not only did NASA put all its space projects on indefinite hold but the Agency shelved, then abandoned completely, its plans to build the second launch site out in California. "It was a traumatic time for everyone," says Wilkinson. "The engineering staff at Goddard had already built most of the COBE hardware."

The explosion of Challenger on 28 January 1986 put an end to plans to launch COBE *on the Space Shuttle.*

But the team's hopes were not entirely dashed. After Challenger, the US government made a decision to get things moving again as quickly as possible and demonstrate that NASA was still capable of launching satellites successfully.

"There was a whole list of satellite projects waiting to go," says Bennett. "Everyone wanted to see their experiments launched."

The task facing NASA was a formidable one. "It had a limited budget," says Bennett. "It had all these projects. And it had some terrible decisions to make. Like what projects do we delay indefinitely? And what ones do we cut out entirely?"

The COBE team went to work looking at alternative launch vehicles. "It was abundantly clear to us that if we didn't get the satellite off the Shuttle, it would never be launched," says Bennett. The team considered hitching a ride on the French Ariane rocket. They even considered talking to the Russians. "Back then, that wasn't nearly as easy to suggest as it is today," says Bennett.

But finally they came back to their original choice—the trusty Delta rocket. It was ironic. After originally designing COBE for a Delta launch then redesigning it for launch by the Shuttle, the team began redesigning it for a Delta once again. It was enough to make anyone despair.

By the end of 1986, Mather and the team had formulated a new plan. It involved halving the weight of the satellite and making all sorts of things on the spacecraft fold up so it would squeeze into the shroud on top of a Delta rocket. "We went to NASA headquarters and said we could rebuild COBE and launch it on a Delta in a short time," says Bennett. "Of course, lots of other projects were coming in saying exactly the same thing. It was a mad scramble."

Getting the weight down was the biggest challenge facing the COBE team. "We had a 10,500-pound spacecraft designed for the Shuttle and the Delta rocket couldn't launch more than 5,000 pounds!" says Bennett.

That it was possible to readapt COBE at all for a Delta launch was something of a miracle. "Fortunately for us, NASA chose COBE as its number one priority, along with the Hubble Space Telescope," says Bennett. "The Agency considered the science solid and genuinely exciting. The buzzword was that it was sexy, and sure to capture the public's imagination. The other thing was that they believed we could do it soon."

Charles Bennett, one of the scientists who worked day-to-day at NASA's Goddard Space Flight Center to ready COBE for its 1989 launch.

At the beginning of 1987 NASA gave COBE the green light, on one condition—that the satellite be ready to go in two years. "Two years is not a lot of time to redesign, rebuild and retest a spacecraft," says Bennett.

The lights burned late into the night

The COBE team was elated. The only doubt that remained was whether all the changes to the satellite could be made in time. "The Goddard engineers did a sensational job of converting the satellite," says Wilkinson. "Without them COBE would have died."

The satellite's own propulsion system was no longer needed to boost the satellite from the Shuttle's cargo bay. It could immediately be dispensed with. "That took 2,000 pounds off at a stroke," says Bennett.

But getting the rest of the weight off was a major problem. Luckily, the giant 600-liter vacuum flask that contained the liquid helium and the three sensitive instruments could just squeeze into the shroud of the Delta. And the boxes containing all the electronics still fitted. But the entire spacecraft skeleton had to be completely remade.

Someone had a bright idea and realized that the old spacecraft structure—basically a large piece of metal for bolting things onto—might come in useful somewhere else. It was sold to another project that was planning to use the Shuttle.

Some things now had to be built folded ready to open up in space. For instance, COBE needed an upside-down umbrella hanging beneath it to hold back the torrent of heat, radar and television signals streaming up from the Earth. This "ground shield" had to be deployable in orbit.

But using such a "deployable" in space is a very risky business. "If it doesn't open, you've had it," says Bennett. "You just can't go up into space and fix it."[4]

For two and a half years, rebuilding the satellite was a major activity at Goddard. Engineers and technicians worked double shifts and weekends to convert the spacecraft, and the lights at Greenbelt often burned late into the night. "Everyone felt a great sense of pride in the project," says Bennett.

Though only about fifty people followed the project from beginning to end, more than a thousand contributed in some way to COBE. "You need to decide what kind of glue to stick this bit to that bit," says Bennett. "Well, some guy is an expert on glue, so he plays a part."

The attention paid to the minutest detail of COBE was phenomenal. "I'll never forget this meeting we had that lasted for two hours," says Bennett. "We were discussing a bolt—a single bolt.

---

[4]At the time this book went to press, NASA's Galileo spaceprobe is orbiting Jupiter unable to send back pictures at a reasonable rate because its deployable communications dish has remained stubbornly closed!

How long should it be? How many turns per inch? What way should it turn?"

"Before I worked on COBE, I always thought people who built space experiments spent too much time worrying about ridiculous details," says Bennett. "But the rocket cost $60 million. You can't go and launch *n* of those, where *n* is a large number."

"These things are incredibly complicated," says Bennett. "It really is a wonder anything works."

Of crucial importance to the experiment was the cold load. It was this that would be compared with the sky to find out how close it was to a black body. The team ended up with a blackened cone, which they strapped to a metal plate, bolted in turn to the vacuum flask of liquid helium.

A critical factor in COBE's success was the fact that the engineers and scientists on the project talked a lot to each other. Some of the scientists, including Mather and Bennett, were at Goddard the whole time, making day-to-day decisions along with the engineers. "The COBE project was small enough that this sort of interaction was possible," says Bennett. "On big space projects, scientists only write a 'requirements document,' hand it to the engineers and the engineers go away and build the spacecraft."[5]

Another thing that helped COBE was the dedication of scientists like Mather and Bennett. By working on the project to the exclusion of almost all else, they freed others, like Dave Wilkinson, to continue working on their own experiments. Wilkinson and the rest were therefore able to keep at the forefront of a fast-moving field while learning more and more about the pitfalls of doing cosmic background experiments. "We were constantly updating our experience," says Wilkinson. "And COBE was able to take advantage of that."

There were always things the experimenters had not anticipated that swamped the tiny signal from the cosmic background radiation. On one occasion, Wilkinson and a graduate student flew a balloon in which a tiny metal switch scuppered the entire

[5]This may have contributed to the problems of the much bigger Hubble Space Telescope, which, in 1988, was launched into orbit with the telescopic equivalent of a squint!

experiment. As the balloon drifted on the winds, the switch cut through the Earth's magnetic field and developed a spurious electrical current. COBE's switches were magnetically shielded to avoid such a problem.

For Wilkinson, the balloon experiments were a breath of fresh air after the frustrations and compromises of COBE. He could be in "complete control." "Dave kept himself aloof from daily involvement in COBE," says Mather. "He came to our meetings several times a year and helped think about what we should do but he did not try to do it." For people like Mather, however, who were deeply involved, COBE was a constant effort.

Shake, rattle and roll

When everything was built, it was time to test it. The equipment has to be able to survive not only the rigors of launch but also the harsh environment of space.

Rocket launches are incredibly violent events. Putting a payload on a rocket is rather like placing a bomb under it. Delta rockets have "accelerometers" at different points, so engineers know exactly how each part of the rocket shakes during launch.

At Goddard there is a building with a table that can shake equipment just as if it were on a real rocket. "You have to watch through a thick window," says Bennett. "You can't be in the same room—the noise would destroy your ears."

The team built fake instruments and bolted them to the test table. They shook the table, but harder than necessary—just to be sure. When nothing dropped off, they put the real instruments on the table and started praying. "It's a scary thing to watch," says Bennett. "You work on these incredibly delicate instruments and then you shake the hell out of them."

The instruments passed their test. It was time to shake the entire spacecraft. Some of the people on the team could hardly watch. They stood around horror-struck, biting their lips, convinced something would go wrong. "You wouldn't believe anything could work after being shaken like that," says Bennett. "If you shook your TV set like that, it would be a pile of junk."

COBE passed with flying colors. "The shake test was a big thing," says Bennett. "The redesigned spacecraft was particularly vulnerable to vibration because it was lighter than we originally planned."

The second big test was the thermal test, which took place in Goddard's Solar Environmental Simulator. The spacecraft was put in a vacuum chamber to mimic space and something hot and bright was shone on it to simulate the Sun. The instruments had not only to survive this baking but also to work well.

The thermal test lasted a whole month. "It was a month when a lot of us didn't get much sleep," says Bennett. Ahead of time, he had drawn up the schedule of tests on the smoothness experiment. It was necessary, for instance, to check how well the satellite's batteries worked and whether its attitude control system functioned as expected.

The COBE team zipped through the test schedule, but from time to time someone would get a result they did not understand, so they would need more time. Bennett was constantly forced to revise the schedule. "You don't go back in once you come out of the Solar Environmental Simulator," says Bennett. "So you have to do your analysis quickly . . . I can't explain how much this business wears you down. Afterwards, you feel utterly drained."

Another thing that had to be done was to make sure COBE's instruments were not "talking to each other." The satellite had to have a radio transmitter on board in order to transmit data back down to the Earth. "We had these extraordinarily sensitive instruments and a several-watt transmitter sitting right in the middle!" says Wilkinson. But none of the transmitter power seemed to leak into the instruments. All was okay.

COBE eventually passed all of its tests with no problems. But the gods had not finished with the nerves of the COBE team. Several months before the planned launch, the team hit problems. The mechanical arm which brought the cold load in and out of the FIRAS spectrometer's trumpet-shaped horn would not stay in place.

The giant vacuum flask was already filled with super-cool liquid helium. Nevertheless, there was nothing for it but to take the lid off, take the cold load out, install a new flexible electrical cable, put it back together and cool it all down again. "It cost us several months," says Mather.

Oh no, it's blown up!

Finally, after two and a half years of nonstop activity, everything was ready. COBE had taken a mammoth 1,600 person-years to build and it had cost a total of $60 million.

Folded up, the satellite was drum-shaped—about six feet in diameter and twelve feet tall. In space, with its solar panels completely unfurled, it would span about twenty feet.

The launch was set for 18 November 1989. The night before, most of the team flew out to Vandenberg Air Force Base, one hundred miles north of Los Angeles. "They got us up at 3 a.m. or some ungodly hour like that and put us onto buses," says Bennett. "It was freezing. I remember trying to get warm on the bus."

When the buses put them down in a field about a mile from the launch pad, dawn was still some time away. "The launch wasn't scheduled for an hour but they'd got us there early," says Bennett. "Of course, there was no guarantee the launch would not be delayed."

It was a large gathering in the field and there was great excitement. Dave Wilkinson was there. He had been in at the very beginning. He was the only experimenter to span the entire history of the cosmic background measurements.

Standing beside him was his "secret weapon." "I think my dad was the only one sad to see COBE come along," says Wilkinson. "It meant we stopped our balloon campaigns, so he could no longer come and help."

Also waiting in the crowd, stamping their feet to keep warm, were Ralph Alpher and Robert Herman. Mather had made a special point of inviting them. Now everyone recognized their prescience in predicting the Big Bang radiation back in 1948.

"That was the most nervous time of all," says Bennett. "Waiting for the launch." Delta rockets are just about the most successful of all launch vehicles, but even they fail occasionally. "Years of our lives had gone into this thing," says Bennett. "It really was all or nothing."

"We had our fingers crossed and crossed again," says Wilkinson. "There were so many things that could go wrong. There were 600 liters of liquid helium on that satellite. It had to get into a high

The launch of the Cosmic Background Explorer Satellite (COBE) *from Vandenberg Air Force Base in California on 18 November 1989. In orbit, 560 miles above the Earth, the 5000-pound spacecraft opened its microwave eyes to the Universe and began observing with its three instruments: the differential microwave radiometer, the far-infrared absolute spectrophotometer and the diffuse infrared background experiment.*

and difficult orbit. It had to be oriented, spun up. The cover had to come off, the ground shield had to pop out . . ."

Dawn was breaking when an extraordinarily bright light flared low in the clear sky. Bennett drew in a sharp breath. He had expected to hear a loud sound first. "My first thought was, oh no, it's blown up!" But everything was fine. "I suppose the people who see lots of these things know it's quite normal."

To the relief of everyone, the blinding light began to climb steadily up into the sky. After fifteen years of effort, COBE was at last on its way. "Seeing it streaking across the sky—that was a beautiful sight," says Bennett.

He did not wait around. As the rocket faded into invisibility, he dashed to a car. "I had to get back to Goddard," he says. "I was the one responsible for getting the smoothness experiment turned on and checked out."

Bennett sped through the desert. "It was funny. I was driving to Los Angeles and I was hearing news reports on the car radio about how COBE was doing. It was my main source of information. The radio reported a successful launch. Then, just before I arrived at the airport, I heard there had been a successful orbit injection, and that the solar panels were working."

From a pay phone at the airport Bennett called Goddard. When he got through to the control room, they told him everything was working well. COBE had reached its orbit 600 miles above the Earth. It was now circling the Earth fourteen times a day, a tiny, drifting star, brightening and fading every 72 seconds as it turned on its axis. It could be seen in the night sky, going from south to north a little after sunset, or from north to south a little before dawn.

COBE awakened, opening its eyes to the microwave Universe.

11

THE NINE-MINUTE SPECTRUM

COBE gets a standing ovation

When John Mather entered the auditorium he was stunned by the sight that greeted him. He had expected about fifty people to turn up for his talk. Instead, it was standing room only and more than a thousand had packed into the lecture hall.

It was 13 January 1990 and COBE had been up in space just seven weeks. The American Astronomical Society was meeting in Crystal City, Virginia, and Mather had come to present COBE's first result—a spectrum of the microwave background based on just nine minutes of looking at the sky.

Mather was determined to remain calm. He launched into the five-minute talk he had prepared, explaining the purpose of the experiment, and proceeding to describe it. Finally, he put a transparency onto an overhead projector so that its image was thrown onto a large screen.

"Here is our spectrum," he said. "The little boxes are the points we measured and here is the black body curve going through them. As you can see, all our points lie on the curve."

"At first you could hear a pin drop in that hall," says Bruce Partridge. "Then there were murmurings in the audience. Next people began to applaud. Then they got to their feet, clapping wildly, enthusiastically."

"I've never seen anything like it at a scientific meeting," says Charles Bennett. "Not before or since."

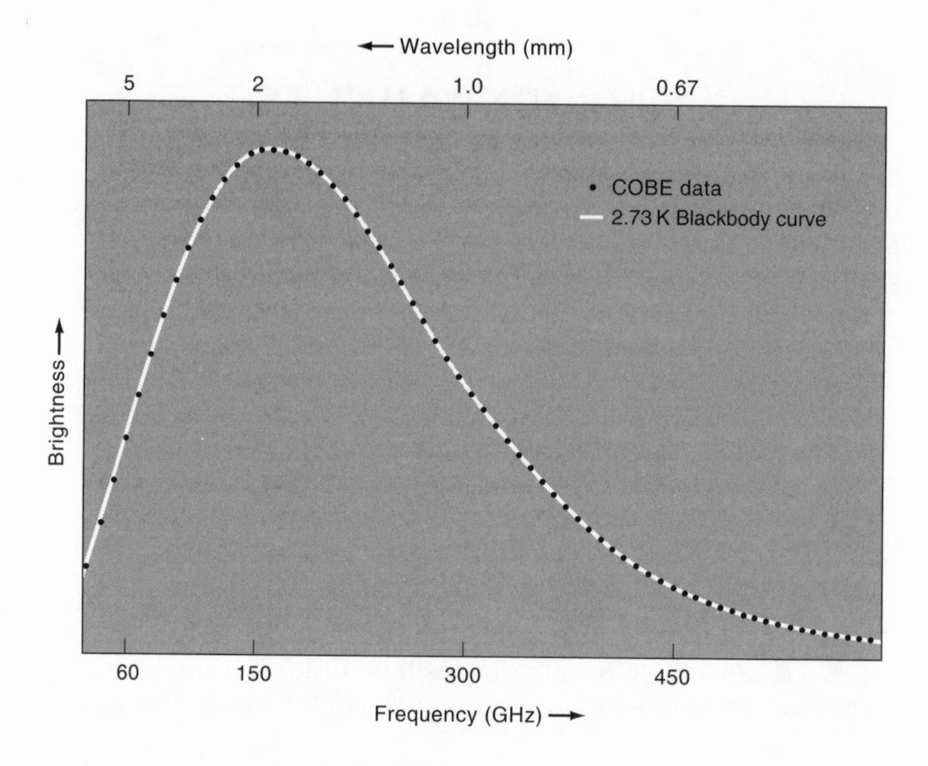

The spectrum of the cosmic background radiation between a wavelength of 0.1 and 10 millimeters, as measured by COBE. *The curve is the spectrum of a black body at a temperature of 2.73 degrees above absolute zero. No data point deviates from the curve by more than 0.25 per cent.*

Up there on the screen was the most perfect black body spectrum anyone had ever seen. Not a single measured point deviated by more than 1 per cent from the mesmerizing curve drawn through them.

"It was a wonderful moment," says Partridge. "The spectrum was absolutely spectacular. There had been rumors that it was going to be impressive but the COBE team had been very good at keeping it a secret."

Mather's immediate reaction to the audience's applause was not pleasure but embarrassment. "I was afraid they were clapping for me," he says. "I wanted to tell them I wasn't the one that did this thing. COBE was a team effort. I played a part but thousands of other people worked on it day and night. They left behind their families just to do it."

But Mather need not have worried. The people were not cheering for him alone. They were applauding a wonderful experiment. They were cheering because no one in that lecture hall had ever seen such perfection emerge from an experiment. Nature was simply not like that. It was messy.

COBE had seen to the very heart of things. It had stripped away all the bewildering complexity of the Universe. And there at the beginning of time was breathtaking simplicity—more beautiful than anyone had dared imagine.

"A lot of that cheering was relief," says Mather. "The scientists were relieved that the Universe was the way everyone had hoped."

There was no sign in the spectrum of the bump found by the Berkeley-Nagoya team. "It seemed that every issue of the *Astrophysical Journal* had three papers speculating on what caused it," says Bennett. "But none of that complicated stuff happened!"

There could not have been a large release of light radiation into the Universe from the decay of microscopic particles or the explosion of an early generation of stars. Almost all of the cosmic background radiation had come straight from the Big Bang.[1]

The early Universe could have been complicated. Its temperature and other properties might have varied wildly from place to place. But they didn't. The early Universe was unbelievably simple. All you needed to know was one number—its temperature—and you knew everything there was to know about it.

Not everyone who was anyone in background work was at the meeting at Crystal City. Dave Wilkinson, for instance, was back at Princeton giving a simultaneous talk on the COBE spectrum. Another notable absentee was Robert Wilson.

The irony was that the co-discoverer of the Big Bang radiation had attended the Crystal City meeting but had decided to go home

---

[1]In fact, COBE would eventually find that the background spectrum differed by less than 0.03 per cent from a perfect black body at a temperature of 2.726 degrees above absolute zero. This implies that 99.97 per cent of the cosmic background energy was released within one year of the Big Bang.

a day early. And nobody on the COBE team had thought to tip him off! When Wilson finally saw the spectrum, he was bowled over by it like everyone else. "It was just marvellous," he says. "I never believed I'd see a spectrum that good. To my mind, it puts an end to the argument about whether this is really from the early Universe or not."

I know a secret

COBE had actually beamed down the spectrum in early December, shortly after the satellite was launched. But the COBE team had kept it a secret. "The pressure on these guys was tremendous," says Partridge. "Everyone knew that if everything worked, once the probe was up and the cover was off the instruments, they'd know within ten minutes what the spectrum was like."

The reason the COBE team kept the spectrum under wraps was that they had an agreement. No one was to talk about any result until everyone was good and ready. This would enable the team to check and recheck a result to make absolutely sure there was no mistake. There would also be time to prepare a rigorous scientific paper before any announcement.

Dave Wilkinson remembers his first sight of the spectrum. It was on a computer screen at Princeton. Ed Cheng, the team member who had generated the spectrum from the raw data, had sent it to him by electronic mail.

"Seeing that spectrum after twenty-five years of knocking off one point at a time was just thrilling," says Wilkinson. "Each of those points had taken a graduate student's thesis.

"Everything on the satellite worked perfectly," says Wilkinson. "After all our bitter experience with balloons. It was just amazing. That complicated thing actually worked!"

Wilkinson's office at Princeton is next door to those of Peebles and Dicke. But because of the team's publication policy he was unable to show either of them the amazing spectrum. For nearly six weeks he drank coffee with Peebles and Dicke without ever spilling the beans.

"I finally showed it to Jim a few days before the official announcement," says Wilkinson. Peebles was not totally surprised at the spectrum Wilkinson showed him. "Dave was walking around with an 'Oops, I swallowed a canary' grin, so I could tell that it looked awfully good."

But though he expected to see something good, Peebles was still not prepared for the sight of such a perfect spectrum. "Dave had been carrying the spectrum about in his pocket for some time," he says. "When he finally got it out and showed it to me, it had all the drama of 'Take a look at this!' It was one of those stunning moments in your life you remember forever.

"The COBE team kept it under cover until they were absolutely sure of the result," he says. "That shows a degree of care that you don't normally see with scientists. Usually, they are in a hurry to get into print."

Peebles admits he never expected to see such a perfect spectrum. "In the real world, when you measure any quantity in nature, there are always errors—the measurements 'scatter' about the real value," he says. "The stunning thing with the spectrum was that the scatter was so small."

Wilkinson had also been surprised when he first saw the spectrum that it was so perfect. "A lot of us were expecting to see the Berkeley-Nagoya distortion," he says. He had grilled Andrew Lange, who had worked with Paul Richards at Berkeley, but had been unable to pinpoint anything the team had done wrong.

"Richards and his team were very careful," says Partridge. "But they were simply trapped by nature."

"It was very hard to create that distortion," says Wilkinson. "We knew that if it was right we'd need to invent some new physics or put a fairly dramatic chapter into the story of the Universe."

Peebles never believed in the Berkeley-Nagoya distortion, something he is proud of. At a meeting on the microwave background just before COBE, he remembers that people discussed the distortion at length. "But none of the theorists at the meeting had a convincing explanation for that effect," says Peebles. "This makes me feel good because the effect wasn't there after all!"

Few thought that all this hard theorizing was wasted, though. "Their result generated a lot of thinking about what could cause the distortions," says Wilkinson. "It was a very useful exercise."

Partridge agrees: "It played the same sort of role as the Steady State theory."

Mather had more faith in the experiment—and in nature. "The spectrum was pretty much what I thought it would be," he says. "The cosmic background radiation really dominated in the early Universe. For every particle of matter there were 100 million particles of light. If you're going to make them not be perfect every particle of matter has got to do multiple duty. It's hard to imagine how that would work."

The best black body ever seen

The COBE spectrum was widely referred to as the best black body ever seen in nature. But the COBE team itself was not prepared to go that far. "All COBE did was compare the sky with the best black body we could make," says Wilkinson. "All we proved was that the Universe is the same as our black body."

The fact that it was the same is why the COBE team is so confident in the result. "If there was anything wrong with the experiment, we wouldn't expect the sky to be like the cold load," he says. "It would be an incredible coincidence if the cold load mimicked the sky and *neither* were black bodies!"

If COBE had detected some kind of distortion it would have been another story completely. "It would have been much longer before we told people," says Bennett. "We'd be wondering, is that really in the sky or is there something wrong with the cold load? It's because it's unlikely the sky and the cold load would match by accident that we have a great deal of confidence in the spectrum."

The COBE team continued to measure the spectrum of the Big Bang radiation more and more precisely. "So far we've found no deviation greater than a thirtieth of one per cent," says Wilkinson. The cosmic background is a true black body with a temperature of 2.726 degrees above absolute zero with no deviations greater than a thirtieth of one per cent of the peak.

The man who was a month too late

Although the COBE team was confident that their spectrum was right, what was needed was confirmation by another experiment.

As it happened, the confirmation would come sooner than anyone expected, from Herb Gush, Mark Halpern and Ed Wishnow at the University of British Columbia in Vancouver.

"Gush is the unsung hero of cosmic background work," says Bruce Partridge. "For years he's worked on a shoestring budget with just a handful of people."

As we have seen, during the 1970s Gush developed the technique of launching cosmic background experiments on sounding rockets. Sounding rockets basically roar up a few hundred miles then plummet back down as soon as their fuel is used up. For a few minutes the instruments on board get to take a quick peek at the Universe from the very edge of space. In principle, when they are above the Earth's atmosphere they can do better than instruments on a balloon that drifts on the winds for ten hours.

Gush pioneered measurements of the spectrum from rockets. His first flights were in the early 1970s but they were plagued by problems, the most serious of which were caused by the rocket exhaust. Rockets are very messy beasts, and all sorts of complicated molecules spew from their exhausts. "Unless you are very careful, you end up looking at the Universe through a thick cloud of smoke," says Wilkinson.

Gush thought he had the exhaust problem solved. Along with his instrument package he included a sort of "ejector seat." It was supposed to blow the experiment clear of the rocket when the right altitude was reached. But things did not go as hoped.

On Gush's fourth rocket flight, in 1978, the ejection mechanism proved too feeble. "The payload blasted free of the rocket all right," says Gush. "But as it sailed on, the rocket overtook it, still burning the last of its fuel." During the seven minutes the experiment was above the atmosphere, it observed the background radiation through a veil of shimmering exhaust fumes.

A spectrum was radioed down to the ground from a height of 200 miles. It was like a black body for the most part. But at millimeter wavelengths there was a large bump. Was the bump really in the background radiation or did it come from the glowing rocket exhaust? It was impossible to tell.

In 1980, when Gush started designing his fifth rocket experiment, more bad luck and frustration were just around the corner. Until now, he had been firing his rockets from a launch pad at

Herb Gush, the unsung hero of cosmic background work, with the parts recovered from one of his rocket experiments. Gush measured a perfect black body for the cosmic background radiation only weeks after COBE *had cleaned up the field. If there was a Nobel Prize for persistence, Gush would have won it.*

Churchill, Manitoba. The Canadian government ran the facility jointly with the Americans, but in the early 1980s decided to pull its money out.

It would be nearly a decade before Gush would fly a rocket again. When he did, it would not be from Canada at all but from the desert of southern New Mexico.

In September 1989, two months before COBE was due for launch, Gush, Halpern and Wishnow were almost ready to go. First, though, they needed to be certain their instrument package would survive the violent vibration of a rocket launch. They took it to Bristol Aerospace in Winnipeg for a "shake test." It failed.

"Only later did we find out that the engineers at Bristol Aerospace had shaken the instrument package too vigorously!" says Gush. Some things had broken loose. There was nothing to do but go back to Vancouver and start repairing the damage. "The extra work cost us five months," says Gush. While the team worked in their lab, COBE was launched and began observing the microwave background.

Finally, in January 1990, Gush was ready to launch. He took the experiment down to White Sands missile range in New Mexico, a facility run jointly by the US navy and army.

On the launch pad, the two-stage rocket stood more than forty feet high, glistening in the morning light. Gush's instruments were crammed into the nose cone, a cylindrical space just three feet high and seventeen inches in diameter.

As the countdown began, Gush sat in an underground bunker close to the launch tower. It was designed to provide protection if the rocket exploded and burning metal and fuel rained down from the sky.

The countdown reached zero and the rocket whooshed into the blue New Mexico sky on a column of flame. Minutes later, it reached an altitude of 200 miles and the instruments were ejected successfully. Sensitive detectors, cooled by liquid helium to just a third of a degree above absolute zero, came alive as the radiation from the Big Bang poured in.

Everything worked perfectly. After twenty years of problematic experiments, Gush had finally done it.

On his way back from the rocket site, Mark Halpern stopped off in Aspen, Colorado, where a meeting on the microwave background was in progress. It was just a couple of weeks after Mather had received his standing ovation in Crystal City.

"Halpern brought with him a beautiful black body spectrum," says Wilkinson. "It was stunning." Gush's team had achieved what hundreds of other experiments had tried and failed to do ever since Penzias and Wilson discovered the background radiation in 1965. And he had done it only weeks after COBE had cleaned up the field. "If it hadn't been for COBE, that spectrum would have got a standing ovation," says Wilkinson.

"They tried and tried again and finally they got it right," says Mather.

"My heart goes out to Herb Gush," says Peebles.

"I suppose they knew they had only one more flight, so they were really careful," says Wilkinson.

"These two experiments were running for more than a decade each and yet by coincidence they came to fruition at almost exactly the same time," says Peebles.

"It would have been a dramatic triumph for Herb if he had got the spectrum first," says Peebles. "But then one measurement had to be made before the other. And one had to be the confirmation."

The parallels with Roll and Wilkinson were hard to avoid. In 1965, they, too, had succeeded in making a ground-breaking measurement of the cosmic background—but only after being scooped.

But Gush had done the community proud. "I think the important thing is that it was an almost instantaneous confirmation of the COBE spectrum," says Peebles. Now nobody could really doubt that the radiation from the beginning of time was a perfect black body.

If there was a Nobel Prize for persistence, Herb Gush would have won it.

12

COSMIC RIPPLES

COBE finds the seeds of galaxies

Measuring the spectrum of the microwave background was just one of the goals of COBE. Once that was achieved, all attention focused on the experiment that was surveying the sky for any sign of unevenness in the Big Bang radiation: the Holy Grail of Cosmology, as it was so often called.

Shortly after the satellite's launch, the COBE team had taken a "quick look" at a small portion of the sky. But that had revealed nothing but unbroken blandness.

By April 1990, COBE had surveyed the entire sky. It had confirmed that one half of the sky was marginally brighter than the other because of the motion of our Galaxy through space. But once this effect was ignored there were no other hot spots in the microwave background. The COBE team concluded that 300,000 years after the Big Bang no region of the Universe was denser than any other by more than one part in ten thousand.

It was now nearly thirty years since Partridge and Wilkinson had begun the first serious search for variations in the Big Bang radiation. And no one had found the slightest trace—apart from the distortion due to the motion of our Galaxy. Where was the imprint of lumpiness in the early Universe? Where were the "seeds" from which galaxies such as our Milky Way formed after the Big Bang?

The matter in the aftermath of creation was spread throughout space amazingly smoothly and yet the Universe we live in, replete with stars and galaxies, is remarkably uneven. The smoothness of the cosmic microwave background seemed to be contradicting the very fact that we are here at all!

There were mutterings that perhaps the Big Bang might be wrong. But very few astronomers would go that far. What was wrong was our understanding of galaxy formation. That was something tagged on to the Big Bang theory. It was an important addition, but an addition nonetheless. The Big Bang itself was pretty incontrovertible. After all, no one could deny that the Universe was expanding, and that it was suffused with a universal background of radiation. Both observations strongly indicated that in the distant past a titanic explosion had occurred in the Universe.

Nevertheless, the scientific community was beginning to get nervous. "If COBE gets to one part in a million and still sees the sky completely smooth, Big Bang theories will be in a lot of trouble," said Dave Wilkinson.

The instrument

The instrument on board COBE that was searching for any variation in the background radiation was called the Differential Microwave Radiometer (DMR). It was a direct descendant of the one used by the Princeton and Berkeley groups in the late 1970s, when both had discovered that the microwave background was a fraction of a degree hotter in the direction the Earth was flying through space.

There was nothing very complicated about the Differential Microwave Radiometer. Apart from a bunch of electronics, all it really contained was a pair of microwave horns arranged in a sort of "V" shape. The angle of the "V" was 60 degrees, which meant that the horns pointed at patches of the sky 60 degrees apart. Each patch was roughly fourteen times the apparent diameter of the Moon.

The electronics would compare the signal picked up by each horn. In this way it would be possible to measure the tiny temperature difference between the two patches of sky. In fact, the Differential Microwave Radiometer was so sensitive that it could detect a difference in temperature of only 0.00001 of a degree.

Measuring the temperature difference between two patches of the sky is a long way from making a map of how the temperature changes over the whole sky. But COBE was moving, and that made all the difference. Not only was it spinning on its axis so that the horns saw patches of sky around a circle, but the satellite was orbiting the Earth. The orbit changed gradually in such a way that in the course of a year the twin horns would measure the temperature differences not between two patches of sky but between millions of patches.

The first hint

The Differential Microwave Radiometer completed its first detailed map of the entire sky in December 1991 after a year of observations. Each of its horns had made a staggering 70 million measurements. The COBE team began to look for fluctuations in brightness.

Each of the measurements was like one piece of an enormous jigsaw puzzle. It was only when they were all put together to make a map of the temperature of the whole sky that patterns started to emerge. This was the hard bit, and it could only be done with the aid of a powerful computer.

The first person to see something was Ned Wright. At the time, the computer at Goddard was still crunching methodically through the data. But Wright was impatient. He devised a way to take a quick peek at the data. He made a rough map and took it to the rest of the team. It had hot blobs and cold blobs on it. Was it really a picture of the Universe as it was 300,000 years after the Big Bang?

At first, everyone was cautious. "There were a dozen things other than the background radiation that could have caused that signal," says Wilkinson.

The biggest worry was that the signal was not coming from the microwave background at all but from our Galaxy. The Galaxy is known to glow at microwave wavelengths so the COBE team had to estimate how bright this glow was and subtract it. It was for this reason that they had in fact included not one pair of microwave horns in the DMR but three.

The three pairs operated at different wavelengths: 3.3, 5.7 and 9.5 millimeters. There were two independent receivers at each wavelength, allowing the team to make six maps of the sky. COBE picked up confusing radiation from the Milky Way at all three of these wavelengths. But the Milky Way was brightest at the longest wavelength. The COBE team used these observations to subtract the Galaxy's emission from the maps they made at the two shorter wavelengths.

When the effect of the Galaxy had been subtracted, the team did indeed have a map of the sky that contained bright blobs and cold blobs. They made a color photograph that showed the whole sky with just what COBE had seen. Mauve patches showed bits of the sky that seemed to be hotter than the average and blue showed colder patches.

Some have called this a "baby photo" of the Universe; unfortunately, it is not really a photograph of the Universe 15 billion years ago. The team knew that most of the blobs were not caused by the microwave background but by electrons jiggling about in their highly sensitive detectors.

After all this incredible effort, the team had a map whose features were partly caused by the sky and partly caused by their detectors, and it was impossible to distinguish the effect of one from the other.

But the COBE team did not despair. They had known all along that this would be the case. After all, they were attempting one of the most difficult measurements in science, one that had defied the best efforts of dozens of astronomers in the past quarter of a century.

The only way to be sure there were real hot spots and cold spots was to compare the maps at wavelengths of 3.3 and 5.7 millimeters. The team projected the maps onto the same screen so they could see them superimposed. They then switched them alternately on and off. Disappointingly, most of the blobs changed. If these had been real structures in the Universe they would have stayed in the same place. Since blobs caused by electrons in the detectors would be spread about entirely at random, they would change. The team therefore concluded that what they were seeing was essentially caused by electrons in the detectors.

But the team did not give up here, either. They had never thought it would be easy. With the aid of a computer they carefully analyzed the two maps. What they found was that a significant amount of the structure did appear the same in the maps at 3.3 and 5.7 millimeters. In fact, it was more than you would have expected by mere chance.

The COBE team had at last found evidence of lumps in the early Universe (see Color Plate 2).

Unfortunately, they could not say precisely where they were. It was impossible to point to any single blob and say, that is a real "lump" in the early Universe. Instead, a "statistical" analysis let the team say how large the fluctuations—or ripples—are at different scales, even though they could not produce a map showing exactly where the bright spots were.

The bright spots were typically 30 millionths of a degree hotter than the average temperature. They occurred on all scales from the smallest COBE could detect—fourteen times the apparent diameter of the Moon—up to the largest—one-quarter of the entire sky.

Ironically, the Soviet Relict I experiment had just missed them when it orbited on board the satellite Prognoz-9 in 1983. But even if it had found anything, it is arguable whether anyone would have believed its result. Relict I's detectors operated at only a single wavelength of 8 millimeters and it picked up unwanted radiation from the Earth because it was badly shielded.

But there were still other possible confusing signals that COBE might have picked up. For many months, the team considered every other possibility. But one by one they eliminated them. No individual signal contributed more than a tenth the size of the signal Wright had spotted.

"We argued all spring," says Wilkinson. But by April 1992, the COBE team was as sure as it ever would be that something was lurking in the data. It was time to make an announcement.

The announcement

A press release was drafted and bounced back and forth between the COBE team and the NASA press office. Finally, everyone was satisfied.

The team also decided to issue a photograph along with the press release. It was the one showing the whole sky, with mauve patches showing bits of the sky that seemed to be hotter than average and blue showing colder patches.

The date and venue for the announcement were fixed. It was going to be at the American Physical Society on 24 April 1992. George Smoot was the spokesman, though Ned Wright, Charles Bennett and Al Kogut would also be up on the podium explaining aspects of the experiment.

The lecture hall was unusually packed that day. Already a lot of excitement had been whipped up. The scientists themselves were tense with anticipation. There had been rumors for at least six months that COBE had found something and to some it had seemed that the Big Bang theory was in trouble.

But there was another reason why the hall was unusually packed. Without the COBE team's knowledge, the Lawrence Berkeley Laboratory (managed by the University of California at Berkeley) had put out its own press release in advance of NASA. It had gone to privileged newspapers, which had been fired with excitement about the story.

George Smoot introduced the work in a twenty-minute talk. He presented the result and tried to give the people some idea of what it all meant. Asked by someone in the audience just how important it was, he said: "Well, if you are a religious person it's like seeing the face of God."

The secret of the Universe?

Nobody was prepared for what happened next as the story raced around the world at the speed of light.

At the *Guardian* newspaper in England, science correspondent Tim Radford watched his fellow journalists utterly transformed. "Everyone in the office who knew even a little about science was rushing about like mad things, saying this was the greatest story ever," says Radford.

At first, Radford wasn't convinced. "But when I got to the end of writing the story even I was beginning to get excited."

The story reached the front page of virtually every major newspaper in the world. You could not turn on a television without

ew York Times

Late Edition
New York: Today, cooler, variably cloudy, rain arriving. High 62. Tonight, rain. Low 52. Tomorrow, gray, damp, raw winds. High 56. Yesterday, high 78, low 53. Details, page B14.

NEW YORK, FRIDAY, APRIL 24, 1992 — 50 CENTS

Photographs by George Tice

car that went out of control.

'njuring Many

Political Memo

Why Perot Could Pose a Threat With $100 Million: It's His Own

By R. W. APPLE Jr.
Special to The New York Times

WASHINGTON, April 23 — If he runs for President this fall, which seems more likely with each passing day, Ross Perot says he would be willing to spend up to $100 million of his own money on the effort — much more than the major-party nominees could spend on their own account.

He would be free to do so, whereas the major-party candidates would not, because he would not accept any Federal campaign money, and they would. In its decision in Buckley v. Valeo in 1976, the Supreme Court ruled that candidates could spend as much as they liked as long as they took no money from the Government.

At one time, there were outcries in the press and from the opposition about rich candidates trying to buy office. The Rockefellers and Kennedys have often been the targets of such complaints. But in this age of public suspicion of politicians' motives, voters may prefer rich candidates who pay their own way to poorer ones who have to make commitments to lobbying groups or other special interests to stimulate the contributions they need.

Herb Kohl of Wisconsin capitalized on that sentiment in his 1988 Senate campaign. Formerly the head of a family supermarket chain, he used his own money for a television blitz with the slogan, "Nobody's senator but yours." Mr. Perot might do the same, and like Mr. Kohl he would benefit from the fact that in looks and manner he is a million miles from the popular image of a Wall Street predator. Seeming to acknowledge his rising popularity, Mr. Perot today delivered his sharpest criticism yet of President Bush. [Page A20.]

Only John B. Connally in 1980, of all the candidates who have sought the Presidency since 1976, went the self-financing route, and he did not do well. A single convention delegate cost him $12 million.

Independent's Disadvantages

Mr. Perot would need no delegates, of course, since he would run as an independent. But as such he would enter the race with a number of disadvantages that would offset his ability to spend lavishly on his campaign.

He would have to create a campaign organization from scratch. His rivals, presumably Mr. Bush and Gov. Bill Clinton of Arkansas, will be able to rely on town, city, county, state and national party committees, as well as the operations they have built up during the nominating sea-

Continued on Page A20, Column 3

Pennsylvania Governor Criticizes Process That's Turning to Clinton

By MICHAEL deCOURCY HINDS
Special to The New York Times

PHILADELPHIA, April 23 — With Pennsylvania's primary only days away, the state's Democratic Governor today criticized his party's Presidential primary process — and its likely nominee, Bill Clinton — saying the

Referring to low turnouts in earlier primaries, Governor Casey said, "We've got a tiny minority of Democrats voting for Bill Clinton, and he's winning every race without generating any sparks, any enthusiasm, any mo-

ighly of the Arkan- illence and intelli- People have a tre- out him. He's got a pport."

ratic Party should of having all un- s to the convention remain uncommit- ild consider nomi- didate should Mr. iore support.

ess the character ' Mr. Casey said. rn this around; I e can't, convention ie selection of an

SCIENTISTS REPORT PROFOUND INSIGHT ON HOW TIME BEGAN

'BIG BANG' THEORY BACKED

Discovery of Wrinkles in Space Yields Clue to Development of Gravity and Cosmos

By JOHN NOBLE WILFORD
Special to The New York Times

WASHINGTON, April 23 — In a momentous discovery supporting the "Big Bang" theory for the birth of the universe, astronomers looking back toward the beginning of time have detected broad wrinkles in the fabric of space. Their discovery is the first evidence revealing how an initially smooth cosmos evolved into today's panorama of stars, galaxies and gigantic clusters of galaxies.

The scientists detected faint temperature fluctuations in microwave radiation echoing from the supposed instant of creation. Though the temperatures fluctuate by no more than a hundred-thousandth of a degree, they signal primeval variations in the universe's topography a mere 300,000 years after its explosive birth.

Creation of Gravity

The variations in topography were large enough, scientists said, to create the gravity needed to attract more and more matter into increasingly expansive clumps. These variations had been predicted by theorists but were never observed until now.

The discovery, made by scientists analyzing satellite data, is being hailed as one of the most exciting and important developments in cosmology in this century.

"What we have found is evidence for the birth of the universe and its evolution," said Dr. George Smoot, an astrophysicist at Lawrence Berkeley Laboratory and the University of California at Berkeley, who led the research.

"This is really a breakthrough, a revolution, in our understanding of the early universe," said Dr. Smoot, announcing the findings at a meeting of the American Physical Society here today.

Cataclysmic Explosion

The nature of the matter creating the density "ripples" has not been determined. But the ripples themselves were created immediately by the original Big Bang. There would not have been time, the scientists said, for the texture to have changed under normal physical processes.

The New York Times *announces the discovery of "cosmic ripples" supporting the Big Bang theory.*

hearing that scientists had discovered the secret of the Universe.

This was one story that the journalists could not be accused of over-hyping. It was over-hyped by the scientific community itself. The fires were fueled by famous scientists. And it was impossible to get more famous than the British theoretical physicist Stephen Hawking. When Hawking said of the COBE finding: "It's the greatest discovery of the century—if not of all time," there was no stopping the story.

Britain's *Independent* newspaper ran the story across its front page with the banner headline: "How the Universe began." Exploding out of the page was a graphic showing the entire history of the Universe from the moment of creation to the present day, with the missing step—the birth of galaxies—now filled in by COBE.

As the British astrophysicist George Efstathiou commented, only major disasters and the marriage of Princess Di have generated equivalent media coverage.

"They have found the Holy Grail of Cosmology," intoned Michael Turner of the University of Chicago. "It is a discovery of equal importance to the discovery that the Universe is expanding, or the original discovery of the background radiation," said Hawking in the *Daily Mail*. "It will probably earn those who made it the Nobel Prize."

The COBE team was taken aback. "I was flabbergasted by the media coverage," says Wilkinson. "We had expected to get some media interest—but nothing like this." Robert Wilson was also amazed. "There was more publicity than when Arno and I actually discovered the radiation," he says.

13

THE HYPE AND THE HYSTERIA

How the COBE results became front-page news

So why did the ripples at the beginning of the Universe make such an enormous splash in the world's media? Was COBE's discovery really as important as the British physicist Stephen Hawking claimed it was?

"We took a fair amount of heat from our colleagues in other fields after Hawking claimed it was the discovery of the century," says Jim Peebles. "It was a wonderful thing—but I'll give you the discovery of the year at maximum!"

According to Peebles, COBE's discovery of the seeds of galaxies certainly did not rank as highly as Hubble's discovery that the Universe was expanding nor Penzias and Wilson stumbling on the faint afterglow of creation.

But Hawking was not alone in making over-the-top comments. "Other scientists said, clearly this is going to be a Nobel Prize," says Peebles. "I don't know why they said all these things, except to speculate they were feeding on each other's enthusiasm."

The irony as far as most scientists were concerned was that the perfect black body spectrum measured by COBE was by far the most important result to come from the satellite. It showed the early Universe to be simpler than anyone had hoped. But the spectrum had received little publicity, despite the rapturous standing ovation the scientists had given it.

The spectrum was not only more important scientifically, it was impressive technically as well. To find the hot spots in the cosmic background radiation, COBE's Differential Microwave Radiometer had to be only twice as sensitive as any ground-based experiment. COBE's measurement of the spectrum was, in fact, an astonishing thirty times as good as anything ever achieved before.

But the greatest irony of all was that it would have been an even bigger story if COBE had *not* found hot spots in the fireball radiation. Then galaxy formation would have been a complete mystery and cosmologists would have had to rethink a lot of their ideas about the Big Bang theory!

Making the whole world catch fire

One organization that benefited enormously from all the publicity surrounding COBE was, of course, NASA. "After its problems with the Hubble Space Telescope and the Galileo probe, the Agency desperately needed a success," says Robert Wilson.

"I'm sure NASA wasn't sorry about all the COBE publicity," says Peebles. "But whether they were a part of generating it, I don't know."

Peebles doubts that NASA could have created the media spectacle even if it had wanted to. "I don't know whether anyone is competent enough to have got this thing going the way it went," he says. "Suppose you hear about some marvellous discovery and you want the whole world to catch fire—would you know how to go about it?"

So why did the whole world catch fire? Several things played a part. Probably the most important was the sheer excitement of the scientists. A lot of tension had been built up because COBE had been up in orbit quite a while and had seen no sign of any variation in the microwave background.

"It was two years between the launch of COBE and the smoothness result," says Peebles. "Ample time for tension to build up. That's one reason there were a lot of people at that meeting where the announcement was made."

There had been a lot of speculation within the scientific community. "For at least six months before the ripples were announced, there were rumors that they had been found," says Peebles. "Science journalists would call me up at Princeton and gently probe me for what I knew. Fortunately, it didn't require any discretion on my part because I didn't know any more than they did."

When the announcement finally came, the tension among scientists had reached fever pitch. "There was a tremendous outpouring of relief," says Dave Wilkinson.

"I think people simply fed on each other's enthusiasm," says Peebles. "There was a sort of psychological reinforcement of excitement. It led to all this burst of publicity. That's my theory."

He says that between the launch of COBE and the unveiling of the spectrum—which he considers the more important result—there was only about a month and a half. "There wasn't time for the reinforcement and excitement to build up," he says.

Reports of the death of the Big Bang are premature

But it wasn't all innocent excitement. Some scientists definitely took advantage of the sudden media interest. "They told the press that the COBE result confirmed the Big Bang," says Bennett. "That wasn't entirely true." In fact, the Big Bang theory was never seriously in doubt. The COBE result was just another brick in a pretty solid foundation.

But in the previous year several groups of astronomers studying how galaxies cluster throughout the Universe had found that this clustering was difficult to explain in terms of the standard version of the theory of cold dark matter. For the purposes of our story it is only necessary to know at this stage that the theory of cold dark matter was something tagged onto the Big Bang theory.

"The press misreported the problems with the cold dark matter theory as the Big Bang being wrong," says Bennett. "A lot of people in the scientific community then made a concerted effort to fix this impression by saying that the Big Bang theory was one thing, and the cold dark matter theory something else entirely."

The trouble was no one listened very hard. "But when our result was announced, a lot of scientists took advantage of the opportunity to correct the previously erroneous news reports and reaffirm that the Big Bang theory was still very much alive," says Bennett.

This was why some scientists were keen to shout out that the Big Bang was okay. Not everyone was simply relieved or excited. Some scientists were setting the record straight.

The Berkeley press release

But other things played a part in heating things up. And one thing was certainly the press release sent out by the Lawrence Berkeley Laboratory. "The COBE story was already on the wires the night before the NASA press conference," says Bennett. "So there was already this underflow of media attention, which we had no idea was going on."

"By the time of the NASA announcement, everyone was warmed up to why COBE's discovery was a wonderful thing," says Mather. "It certainly got us a lot of publicity."

At NASA, the Berkeley press release caused consternation. "The Agency had tried its best to be fair to everyone," says Bennett, "but it had journalists coming in and saying, 'How come you gave him this picture and me this picture?' "

The NASA press office did not have any idea what was going on. It wasn't until later that it realized there were *two* press releases—the NASA press release and the one from Berkeley.

"Some favored newspapers, like the *Wall Street Journal*, were given a jump on the rest," says Dave Wilkinson.

"NASA doesn't usually do things this way," says Bennett. "It has a sense of fairness with the media and does not preferentially release things to favored journalists. Berkeley does. I don't fault them at all for their policy. But in this case it conflicted with NASA."

"Because some journalists got a jump on the others," says Wilkinson, "they wrote more detailed and more splashy stories than they would have done if they'd simply gone to the NASA press conference."

"In fact, many reporters wrote their articles without even knowing there was a NASA press release," says Bennett. "There was nothing wrong with Berkeley putting out a release," he continued. "the mistake was not checking it with the rest of the team."

"The Berkeley PR machine is extremely good," says Bruce Partridge.

"George Smoot had gone through a tremendously stressful period analyzing the data," says Wilkinson. "I guess when the Berkeley press office told him they were going to release a little earlier it didn't register with him that it was violating our agreement."

According to Smoot, there was no violation of the NASA deadline. Although the Lawrence Berkeley Laboratory press release was sent out to only five places ahead of time, including the news agency Associated Press, it was embargoed for release on Thursday 23 April, the day of the NASA press conference. Associated Press then sent its own release around the night before, with a similar embargo stamped on it. "To the best of my knowledge nobody broke that embargo," says George Smoot.

Bad feelings

But quite apart from their feeling of betrayal, the COBE team was deeply upset by the content of the Berkeley press release. "It focused undue attention on Berkeley," says Mather.

"There was little mention of the people who did most of the work," says Bennett.

"When I switched on my TV and heard that the Berkeley team did the experiment, it upset me," says Wilkinson. "It was a complete distortion. Most of the work on COBE had been done by the people at Goddard—John Mather and the rest—and the Berkeley press release did not give them due credit."

"John Mather had bent over backwards to give the team credit," says Bruce Partridge.

"A lot of young people who worked very hard on this didn't get mentioned," says Wilkinson. "They were pretty upset."

"John Mather is the guy," says Bennett. "He's very self-effacing. There's a big personality difference between Mather and Smoot."

Smoot had gone to NASA headquarters a month before the NASA press conference to help write up the official release and get it cleared. It was the second NASA announcement concerning the smoothness experiment. Back in Berkeley, he told his boss and his head of lab about the NASA release. Smoot says there was a general feeling at Berkeley that in the first release NASA had given too much credit to Goddard and not enough to Berkeley. A Berkeley release was drawn up. "I insisted on a joint release," says Smoot. "NASA had to get first credit."

The COBE team was completely unprepared for the bad feeling the Berkeley press release caused. Until now, the members had worked together harmoniously. "We had to get this thing resolved—it was splitting the team," says Wilkinson.

"Smoot admitted he'd made a mistake," says Bennett. "He apologized to the team."

"George has done everything he can do to put things right," says Mather.

But in the eyes of the public, George Smoot had become COBE. "It's unfortunate but that's what has happened," says Wilkinson.

Smoot says this sort of thing always happens when the press covers a story. Inevitably, one person ends up under the spotlight. "The first day, the press coverage was pretty even, with quotes from me and Ned Wright and the others who were on the platform at the NASA press conference," says Smoot. "But over the next few days, more and more Smoot quotes got used."

But Smoot believes positive things came out of the Berkeley press release. "It made the whole thing a bigger story," he says.

On this score the COBE team was not entirely innocent. If the "Berkeley business" muddied the waters, the team was guilty of doing the same thing by releasing the photograph they did—the one with the mauve blotches for hot spots and blue blotches for cold spots. The photograph was reproduced in virtually every major newspaper and magazine the world over, and most people who saw it assumed they were really seeing clumps of matter in the Universe 15 billion years ago.

"The picture caught everyone's attention," says Wilkinson, "but it was misleading."

"It was real structure in the early Universe mixed in with instrumental noise,"[1] says Peebles. "It certainly wasn't the face of God!"

---

[1]"Noise" is just a technical name for the random jitterings of electrons inside any material.

George Smoot, one of the prime movers behind COBE *who shot to stardom after the cosmic ripples announcement.*

"We saw this problem coming," says Wilkinson. "In fact, there was a lot of debate about the photograph on the team. Should we or shouldn't we use one at all?"

"We knew that most of what was in the picture was not real," says Bennett. "But the overall feeling was that we should show the picture but be careful to caution people that what they were seeing were the biggest things in the Universe plus a whole lot of noise from the instrument."

According to Bennett, the team dreamed up an analogy that they never used to explain the picture. It involved interference, or "snow," on a television screen. "If you're a long way from a transmitter and you turn on a TV, you get all this snow on the screen," says Bennett. "But, amid all the snow, you can still see the vague outlines of a picture. Well, the picture we released of the early Universe had a lot of snow on it."

One person on the team thought the problem of conveying what the picture showed simply too great. "I advocated not using the picture at all," says Wilkinson. "I knew none of the media would take the time to explain that the picture was half noise and half pattern."

"But there was a feeling on the team that we should show people a picture to get over the idea we were looking at the whole sky," says Bennett. "Unfortunately, the noise part of the caution got dropped along the way."

"Perhaps people on the team were a little naive about how such things get covered in the press," admits Bennett. "You show the picture with the explanation and you know it's the picture that will run and the explanation that will be dropped."

The face of God

But to many on the team, the misleading photograph paled into insignificance compared to George Smoot's face of God comment at the NASA press conference.

"When George came out with that, it was a complete surprise to all of us!" says Bennett.

In advance of the NASA press conference, the COBE team had discussed what should be said. "We didn't exactly go over it word for word," says Bennett. "But we agreed on the general tone, removed scientific jargon, that sort of thing. Nobody mentioned the face of God."

"I made the comment on the spur of the moment," admits Smoot. He says he never intended to connect the COBE discovery directly to God but only to convey to nonscientists some idea of how important it was. He hit on: "If you're religious, it's like seeing God."

"I think that was going a bit far," says Bob Dicke.

"George was trying to get over the enthusiasm and excitement we all felt, which was a very positive thing," says Bennett. "But bringing in the religious connection was a potentially dangerous mistake."

Smoot never expected people to take his comment literally. But that is exactly how some newspapers did take it, giving their readers the distinct impression that in the depths of space the COBE scientists had really found traces of God!

"George has a rather extrovertish personality," says Bennett. "He says things to the press in a way he would not talk to a scientific audience."

A debate ensued about what science could or could not say about God. It was all pretty irrelevant to COBE. It served only to muddy the scientific waters, making it harder for ordinary people to understand what it was the satellite had actually found.

In Britain, the *Daily Telegraph* asked cosmologists and clergymen to comment on the COBE discovery under the headline "Cosmology versus theology." And the spurious religious connection was aired in television debates as well. In the United States, Bennett was asked by NBC to do their morning phone-in show to discuss the religious aspects of the COBE result. "Not on your life!" he told them.

What alarmed a lot of scientists was that claims were being made for science that were not justified. Most scientists agree that science illuminates the "how" of the Universe, but has nothing whatsoever to say about the "why," which is the preserve of religion.

"We've taken a lot of ribbing from fellow scientists for the things George said," says Mather.

Smoot says he is passionately interested in communicating science to the public, something he is actively doing at Berkeley. "If my comment got people interested in cosmology, then that's good, that's positive," he says. "Anyhow it's done now. I can't take it back."

The Indiana Jones of physics

The Berkeley press release and the "face of God" comment helped to make George Smoot's name synonymous with COBE.

Shortly after the "ripples" announcement, Smoot received a phone call from John Brockman, one of the highest-profile literary agents in the publishing world.[2] Brockman, who was on a business

---

[2]The following information is based on an interview with John Brockman by Michael White, published in the *Sunday Times* on 13 December 1992.

trip to Japan, was ringing from a pay phone at Tokyo airport. On the way to the airport, he had noticed a newspaper headline declaring that there had been a breakthrough in our understanding of the Universe. Smoot's name was mentioned prominently.

When Brockman got through to Smoot in California, he reportedly said: "Hey look, something big is happening in the Universe, what's in it for me?" Before Brockman's money ran out, he had got Smoot to agree to write a proposal for a book and fax it to Brockman's New York office so that it would be there when Brockman arrived back in the United States thirteen hours later.

Brockman had struck a chord with Smoot. "Even before the COBE announcement, I was interested in writing a book about cosmology," says Smoot.

Brockman arrived back in New York to find the fax waiting for him. He worked on it for twenty-four hours. Within two days of seeing the newspaper headline in Japan, Brockman had the proposal in the offices of sixty publishers in twelve countries. Within a week, he had auctioned the book in New York, London, Munich, Milan, Barcelona and Paris for the largest deal in the history of science publishing—reportedly in the region of $2 million when all the individual national deals were added up.

Smoot had become a major celebrity. He appeared on chat shows and news programs. Magazines articles were written about him.

On 15 November 1992, George Smoot was featured on the cover of the *Boston Globe* magazine. Inside, staff writer Mitchell Zuckoff called him "the planet's most popular astrophysicist" and portrayed him as a sort of cross between a scientist and a movie star. "If Indiana Jones were a physicist instead of an archaeologist," wrote Zuckoff, "he'd be George Smoot."

Eyes on the Prize

There was loose talk about COBE's achievements deserving the recognition of the Nobel Prize. And who could rule out that possibility? After all, the discovery of the cosmic background radiation by Arno Penzias and Robert Wilson had been deemed worthy of a Nobel Prize in 1978. The Nobel committee was notorious for its caution, and often waited years—even decades—before bestowing its accolade on those who had made an important scientific discovery.[3] But with COBE, there was no need for caution. Both the satellite's major results had been confirmed by experiments from the ground, so there was little chance of the Nobel committee getting egg on its face by backing a discovery that would next year vanish in a puff of smoke.

But if COBE's achievements were deemed worthy of the ultimate accolade, who should get the Prize? The satellite, after all, was very much a team effort—hundreds of people had been involved in the project over the past two decades. The obvious choice would be John Mather—the man who in 1974 conceived the COBE idea. Mather not only pushed the project to its completion but was largely responsible for the most successful instrument on the satellite. But if Mather were to get the Nobel Prize, who should share it with him?

There were many possibilities. But one man had separated himself from the pack. That man was George Smoot.

---

[3]Albert Einstein had to wait sixteen years for recognition, and even when he got the Prize it was not for relativity but for his work on the photoelectric effect.

14

THE UNIVERSE ACCORDING TO COBE

Galaxy formation, dark matter and inflation

Amid all the media hubbub, it was difficult to tell just what COBE had found and impossible to tell what it all meant. Many people who watched TV the night of the "cosmic ripples" announcement or read newspapers were rendered dizzy by the convoluted cosmological explanations. They wondered whether it was mere hype or whether the COBE satellite really had discovered something of great importance.

One thing was for sure. COBE did not unravel the mystery of the Universe, as some newspapers claimed. But the satellite did supply important information that provides a crucial missing link in modern astronomical theory. COBE found that the temperature of the cosmic background radiation differed ever so slightly in different directions. The sky contained hot spots and cold spots, often referred to as ripples. The hot spots were just 30 millionths of a degree hotter than the average temperature of the sky, so it was no wonder it had taken more than a quarter of a century to find them. The motion of the Earth through space created an effect a hundred times bigger.

The hot spots marked regions of the Universe that were marginally less dense than average, whereas the cold spots marked the denser regions, or lumps. The lumps were on an enormous scale—between 100 million light years and 2,500 million light years across. They were the oldest and largest structures in the Universe—the "seeds" of giant clusters of galaxies in today's Universe.

Now at least we could be sure that we existed!

Other explanations for what COBE found

Of course, in interpreting the COBE result, astronomers were making the tacit assumption that the last time the photons of the cosmic background radiation were in contact with matter was when atoms formed 300,000 years after the Big Bang. But what if the photons of the background radiation had interacted with particles of matter during their long journey to the Earth? They might be telling us nothing at all about the lumpiness of matter at the beginning of the Universe.

One way this could have happened was if the Universe had been reheated to thousands of degrees at some time during the past 15 billion years. Electrons would have been freed from atoms so that they could scatter the background photons. The reheating could have been caused by an early generation of stars that blazed brightly at the dawn of time, before any galaxies formed. If there had been such a generation of stars—and there was no evidence for them—then the cosmic background radiation, instead of carrying a snapshot of the Universe 15 billion years ago, might be carrying an imprint of this later era.

But as the COBE team continued to measure the spectrum of the fireball radiation with ever greater precision, this possibility began to look increasingly unlikely. If the Universe had been reheated in the past 15 billion years then this should also show up as a distortion of the fireball spectrum. Instead, the spectrum showed no discernible deviations whatsoever from a perfect black body, strong evidence that it did indeed come directly from the Big Bang.

Some theorists suggested that on their way to the Earth the photons of the cosmic background radiation might have instead been influenced by the gravity of so-called cosmic strings. These bizarre objects have been likened by theorists to the cracks that form in ice as it freezes, except that these "cracks" formed in the fabric of space as it cooled after the Big Bang. Cosmic strings were bits of space that got left behind in a hot dense state as the Universe cooled. Preserved along their length were the conditions of enormous density that prevailed in the first moments of creation. If cosmic strings were scattered about the Universe, then any cosmic background photons passing near one would lose energy pulling themselves free of its intense gravity.

But the idea that cosmic strings had caused the cold spots COBE had seen in the sky had its problems. If such bizarre objects really exist in the Universe, then they ought to distort the images of distant galaxies. So far, astronomers have not seen such an effect.

But even if the cosmic background radiation did come straight from the Big Bang, there were things other than the lumpiness of matter that could have left their mark on it. For instance, the hot and cold spots could have been caused by gravitational waves—ripples in the very fabric of space—created by violent events in the first split second after the Big Bang.

The American physicist Craig Hogan has even suggested that variation in the temperature of the sky might be caused by astronomical objects at very large distances. If there were a lot of them, their light could add up and produce the lumpy signal seen by COBE. But Bennett believes this idea can be ruled out. "We've done correlations with data bases of distant extragalactic objects and we don't find you can explain most of the signal that way," he says.

Bennett admits there is no unique way to explain the COBE result. But he thinks the alternative ideas are unlikely. "The COBE team feels that the simplest explanation for what we are seeing is lumps of matter in the early Universe," says Bennett.

The invisible Universe

The implications of the COBE result go far beyond galaxy formation. For one thing, the result bolsters the theory that most of the Universe is made of invisible, or "dark," matter. The reason for this is that the lumps of matter COBE had found in the early Universe were simply not big enough to suck in all the matter to make galaxies or clusters of galaxies in the 15 billion years the Universe has existed. They needed help—from a lot of dark matter.

The strange notion that most of the matter in the Universe is invisible had its origin back in the 1930s. It was then that the Swiss-American astronomer Fritz Zwicky discovered something

Fritz Zwicky, the Swiss-American astronomer who first guessed that the visible Universe was merely the tip of an iceberg.

peculiar when measuring how fast galaxies were moving inside clusters of galaxies. Zwicky found that most galaxies were moving faster than they should. Long ago, they ought to have broken free of the gravitational clutches of their parent clusters and sailed off into the wider Universe.

The only explanation that Zwicky could offer for why they had not sailed off was that the parent clusters contained more matter than he could see with his telescope. It was the combined gravity of this hidden, or "dark," matter, said Zwicky, that was keeping the visible galaxies prisoners within their clusters.

Zwicky was a little ahead of his time in coming to this conclusion, and it took the rest of the astronomical community several decades to catch up. But by the 1980s, it was abundantly clear to everyone that Zwicky's anomaly could not be swept under the carpet.

The evidence for dark matter was incontrovertible. Everywhere astronomers looked in the Universe they found evidence of its ghostly presence. Even our own Milky Way was found to be embedded in a massive spherical cloud of dark matter, which greatly outweighed all of its visible stars. Astronomers now believe that between 90 and 99 per cent of the mass of the Universe is in the form of "nonluminous" dark matter, detectable only because its gravity influences the trajectories of the visible stars and galaxies.

This disconcerting discovery has put astronomers in a hugely embarrassing position. Everything they have been studying with their telescopes these past three hundred years turns out to be only a tiny fraction of all there is. Ordinary matter, which scientists have dedicated themselves to understanding—the stuff of planets and stars and the atoms of our own bodies—is no more than a minor contaminant in the Universe!

And what is even more embarrassing to astronomers is that they have no good idea what the dark matter is made of. There has been no shortage of suggestions. For instance, it could be made of collapsed stars like black holes[1] or even of brown dwarfs, failed

---

[1]A black hole is left when a massive star shrinks under its own gravity. In the process, its gravity becomes so strong that even light cannot escape—hence a black hole's blackness.

stars that are so faint we could easily miss them with our telescopes. Then again, the dark matter could be made of exotic microscopic particles so far undiscovered. Physicists have given these hypothetical particles names like axions, photinos and gravitinos, but nobody is hugely confident that any of them really exists.[2]

But whatever the true identity of the dark matter, COBE's discovery of lumps in the early Universe only emphasized that there had to be an awful lot of it around. Without it, clusters of galaxies simply could not form.

According to the accepted theory of galaxy formation, regions of the early Universe where the matter was slightly denser than elsewhere naturally grew at the expense of other regions. They pulled in more and more matter because their gravity was stronger than the gravity of their surroundings. But the trouble with the lumps that COBE found was that they were only marginally denser than their surroundings. It would take the gravity of such lumps longer than the 15-billion-year history of the Universe to pull in enough matter to make a cluster of galaxies.

But if the Universe contains a lot of dark matter, the dark matter would have speeded things up because it would have curdled into clumps much sooner after the Big Bang. The reason for this is that it was unaffected by radiation. It neither emitted light nor absorbed it, nor interacted with light in any other way. This was in marked contrast to ordinary matter, which was constantly being blasted apart by the photons of the fireball radiation.

Each clump of dark matter that formed would have exerted a strong gravitational pull on its surroundings. But ordinary matter would not have fallen into its clutches immediately; the pressure of fireball radiation would have kept it spread out very smoothly. But though it would have been smooth, it would not have been dead smooth. Around the lumps of invisible dark matter, ordinary matter would have been concentrated ever so slightly.

Finally, when atoms formed 300,000 years after the Big Bang, ordinary matter was freed from radiation so that it could begin to

---

[2]At the moment, many physicists are carrying out experiments to look for such particles at the bottom of old mines or in mountain tunnels.

clump. At this time, according to the theory, ordinary matter was denser by about ten parts per million in the vicinity of each clump of dark matter than it was on average in the Universe. This is very close to the density difference measured by COBE for lumps of matter in the early Universe.

Once atoms formed and the Universe became transparent to radiation, there was nothing to keep ordinary matter out of the gravitational clutches of the dark matter. It quickly clumped to form stars and galaxies. With dark matter helping it along, this process of galaxy formation was greatly accelerated. In fact, it could be completed in the time available since the Big Bang.

Hot and cold dark matter

The dark matter we have been talking about so far is known by the theorists as "cold" dark matter. Cold just means it consists of some kind of particles that are moving sluggishly. Such particles can be easily tamed by gravity and tend to clump rather like ordinary matter.

"The cold dark matter model for making galaxies is a beautiful idea," says Jim Peebles. "I can say that because I was one of the people who invented it!"

Although cold dark matter could have helped galaxies to form more quickly, it has a problem. When astronomers simulate the whole process of galaxy formation on a computer, they find that they end up with clusters of galaxies that are subtly different from those they observe with their telescopes.

In recent years, astronomers have found that the Universe contains structures on scales bigger than they ever expected—great chains and walls of galaxies. Although cold dark matter is good for explaining some of the relatively small structures of the Universe, such as galaxies and clusters of galaxies, it is not good at making these large structures.

It isn't absolutely clear that cold dark matter cannot explain these gigantic structures because there are uncertainties in the observations and the theory. But even some of the proponents of cold dark matter—including Peebles, who invented the idea—have begun to worry just a little. "The cold dark matter model of galaxy formation is in deep trouble," he says.

But there is another type of dark matter that theorists can envisage, and some have invoked it to help explain the way galaxies cluster. The other type is known as "hot" dark matter. The particles that make this up would have come out of the Big Bang moving very fast—in fact, close to the speed of light. It is difficult for gravity to tame such particles so they would be spread far more evenly throughout the Universe than particles of cold dark matter.

The gravity of hot dark matter would therefore tend to keep ordinary matter spread out. In contrast with cold dark matter, which is good at making the small-scale structures, hot dark matter would make the large-scale structures.

Some theorists have begun to claim that both types of dark matter are needed in the Universe. Of course, nobody said the Universe had to be simple and that there had to be just one type of dark matter.

Cynics would counter that the theorists, rather than throwing out their theory of galaxy formation because it conflicts with what we observe, are simply trying to patch it up by putting a band aid on it. But then they have little choice. The great white hope of galaxy formation—cold dark matter—is on the ropes. It was reeling before COBE, but the satellite dealt it a further blow from which it may never recover.

The latest astronomical discovery, made by astronomers in the USA and Britain, only underlines that galaxy formation is still a puzzle. Paul Francis of the University of Arizona's Steward Observatory and Paul Hewett of the Institute of Astronomy in Cambridge believe they have found the two most distant galaxy "superclusters"[3] ever seen.

Francis and Hewett have not seen the superclusters directly but have found evidence for them in the spectra of two distant quasars. Nobody really knows what quasars are but the important thing, as far as the latest work is concerned, is that quasars are so bright that they can be easily seen at the edge of the Universe. Their light, on its long journey to the Earth, passes through many clouds

[3]A supercluster is a great collection of galaxy clusters.

of gas and galaxies that are too far away to be seen even with our biggest telescopes. But these intervening objects absorb some of the quasars' light, and so leave their indelible imprint on the quasars' spectra.

Francis and Hewett have found that the spectra of their two quasars contain "absorption lines," which tell of two giant superclusters of galaxies floating in space between us and the quasars. The superclusters existed barely two billion years after the Big Bang. But how could that be? In the cold dark matter model of galaxy formation, large structures in the Universe slowly assemble from smaller ones, and this takes time. As Francis and Hewett point out, two billion years is not enough time to make the superclusters they have found.

Mystery is heaped on mystery. After COBE, there are more questions about galaxy formation than there are answers. But that is so often the way with scientific discovery.

The bang before the big one

Apart from bolstering the theory of dark matter, the hot spots found by COBE were widely claimed to have proved another esoteric theory of the early Universe known as "inflation." The theory predicts that hot spots should range over all sizes and that they should have the same temperature whatever their size—precisely what COBE found. However, inflation is not unique in making this prediction and people went just a little overboard in their claims for the theory. They desperately wanted inflation to be true. In the words of Jim Peebles: "If inflation is wrong, God missed a good trick!"

The reason the theory is so attractive to astronomers is that it seems to solve at least one major cosmological puzzle and at the same time explain just what the Big Bang was. "Inflation is a beautiful idea," says Peebles. "However, there are many other beautiful ideas that nature has decided not to use so we shouldn't complain too much if it's wrong."

According to inflation, which was proposed in 1980 by Alan Guth of MIT, there was an era before the Big Bang. Although this lasted for only a split second, the Universe nevertheless managed

Alan Guth, the theorist who proposed that the Universe "inflated" from a tiny seed of matter and is therefore "the ultimate free lunch."

to undergo an extraordinarily violent expansion or "inflation."

It is almost impossible to convey just how violent the expansion was. Some have likened it to a nuclear explosion compared with the hand grenade of the Big Bang. Others have simply pointed out that during inflation space blew up from a volume smaller than a proton to a volume bigger than the observable Universe, 10^{50} times bigger, where 10^{50} is mathematical shorthand for 1 followed by 50 zeroes!

Inflation was over and done with by the time the Universe was a million-million-million-million-millionth of a second old. Thereafter, the Universe expanded at a much more sedate pace. This sedate expansion was the Big Bang, which until Alan Guth had come along everyone had considered the most violent explosion imaginable.

Inflation is almost impossible to understand unless you are a particle physicist![4] It is an outgrowth of the modern picture of the vacuum. Incredibly, the energy that drove this enormous expansion came from empty space itself. Far from being empty, the vacuum is now known to be seething with "quantum fluctuations," subatomic particles and their antiparticles that are permitted by

[4]So don't worry if what follows makes little sense. You will be in good company!

the Heisenberg uncertainty principle to appear out of nothing and annihilate each other in the blink of an eye.

These "virtual" particles have real and peculiar consequences. Not only do they create a small mass density even in empty space but associated with this is a "negative pressure," a sort of tension in space. In normal circumstances, this pressure is minuscule but when the Universe was about 3×10^{-44} seconds old and the density of both matter and the vacuum a staggering 10^{94} grams per cubic centimeter, a very special "vacuum-like state" was created in which the negative pressure was overwhelming. It manifested itself as a kind of gravitational repulsion. And it was this that provided the enormous "push" that started the expansion of the Universe.

But the vacuum-like state was unstable. Like an atom in an "excited" energy state, it desperately wanted to return to its "ground," or lowest energy, state. After about 3×10^{-35} seconds, the special vacuum decayed into the familiar vacuum we know today.

The consequences of this were dramatic. During its phenomenal expansion, the density and pressure had remained constant—another bizarre property of the special vacuum state. This meant there was an enormous amount of energy available. When the vacuum decayed, the energy went into the creation of ordinary particles moving near the speed of light.[5] Consequently, the Universe became hot, its temperature leaping to a fantastic 10^{27} K. This was the hot Big Bang.

The principal problem that inflation solves is the so-called horizon problem. This is that the temperature of the cosmic background radiation is virtually the same in regions of the sky that could not have been in contact with each other at the epoch of last scattering. They were beyond each other's "horizon." Inflation neatly solves this problem because our entire Universe inflated from a minuscule region of space, and portions that are very far apart now could easily have been very close together in the first split second of the Universe.

The reason scientists were overeager to say that the COBE result proved inflation was correct was that the theory provides a natural

[5]Einstein showed that matter and energy are interchangeable and connected by the famous equation $E = mc^2$, where c represents the speed of light.

way both to create tiny variations in the density of the Universe during the first split second of creation and then to magnify them to the size seen by the satellite.

During the inflationary era particles were constantly being created out of radiation before being destroyed again. These so-called quantum fluctuations meant that the density of matter was ever so slightly different from place to place. The difference, predicted precisely by the laws of particle physics, was so small as to be almost nonexistent. But during the enormous inflation of the Universe, these tiny density differences could have been magnified tremendously. They could have ended up as the lumps of matter that were seen by COBE.

The implication of this is as startling as the idea of inflation itself. If the theory is right, then the huge chains and walls of galaxies seen by COBE, which are more than 100 million light years across, started out in the newborn Universe as tiny quantum fluctuations smaller than the size of an atomic nucleus! There could be no more dramatic connection between the physics of the very small and the very large.

A further implication of inflation is that only a tiny "seed" of matter was needed in the Universe before inflation. The balance of the matter we see around us today was then created from the prodigious energy of the vacuum. This has led to the suggestion that the seed was itself a quantum fluctuation, making the Universe the "ultimate free lunch."

As for what happened before inflation, here we are into even more speculative realms. In some inflationary scenarios, the observable Universe is no more than a bubble inflating in one corner of a greater universe. Elsewhere, forever inaccessible to us, may be an infinity of other expanding bubbles like the froth on a great sea.

Solving the horizon problem

Inflation is not unique in predicting the properties of the lumps of matter seen by COBE. But what is unique about the theory is that it explains in a very natural way one of the deepest puzzles of the

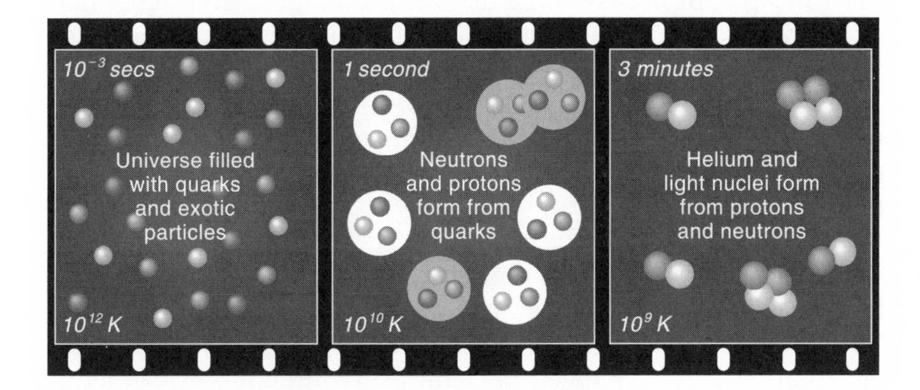

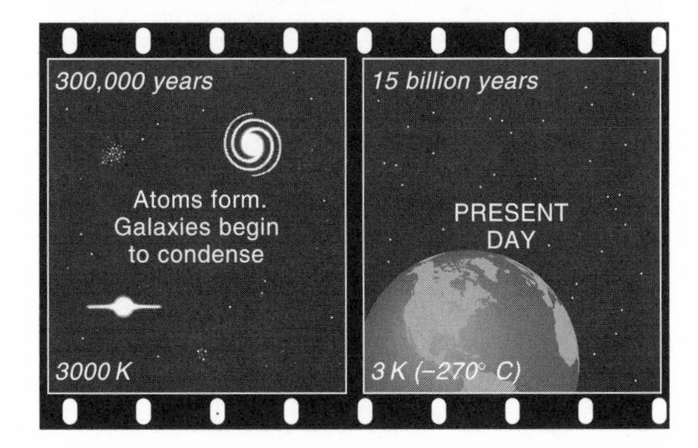

Movie of the Universe: theorists can run the expansion backwards, frame by frame, but the moment of creation still remains shrouded in mystery.

cosmic background radiation: why its temperature is so nearly the same in all directions.

The problem is that the fireball radiation coming from opposite directions in the sky was emitted from regions of the early Universe that could not have been in contact with each other 300,000 years after the Big Bang. However, their temperatures could have kept in step as they cooled *only* if they were in contact.

To see why, imagine two mugs of hot coffee brought into contact. If the first begins cooling marginally faster than the second, then heat will flow into the first mug from the second, and the pair will promptly be brought back to the same temperature.[6] A similar thing will happen if the second mug gets ahead of the first one. The two mugs will cool at the same rate so that at all times they always share the same temperature. On the other hand, if the two mugs are not in direct contact—for instance, if they are in different parts of a room—there will be nothing to stop them from cooling at different rates. If one is in a draft, for example, its temperature could easily drop more quickly than the other.

In the same way, if two regions of the early Universe were to have shared the same temperature as they cooled, heat must have flowed between them. But there is a limit to how fast this could have happened—the speed of light. So two regions could have stayed at the same temperature only if they were close enough for light to have travelled between them in the time since the beginning of the Universe.

And herein lies the problem with the cosmic background radiation. When astronomers look at the fireball radiation coming from opposite sides of the sky, what they are seeing is light emitted by regions that were much further apart than any influence could have travelled in the 300,000 years since the beginning of the Universe. In fact, only regions separated in the sky by less than about four times the apparent diameter of the Moon could possibly have been in touch, and so have any right to share the same temperature.

But if the Universe did indeed go through an inflationary era before the Big Bang, this problem—known as the "horizon prob-

---

[6]Heat always flows from a hot body to a cold body—something physicists have enshrined in the second law of thermodynamics.

lem"—has a very natural solution. Namely, that our Universe inflated from a region smaller than a proton in an atomic nucleus. The region was so small at the time inflation began that light had had plenty of time to cross it since the beginning of the Universe. So regions of the early Universe today seen on opposite sides of the sky were in very close contact before the inflationary era began. They had plenty of time to reach a common temperature.

The hot spots seen by COBE were so large that light could not have crossed them since the beginning of the Universe. This is the strongest evidence that they were imprinted on the Universe well before the time when matter and radiation went their separate ways, 300,000 years after the Big Bang. But this is not proof that they were imprinted in the first split second of the Universe, as inflation requires.

If inflation is right—and, in truth, the COBE result is simply compatible with the idea—then the implications for the Universe we live in are considerable. The region of space we see with our telescopes may be only a vanishingly small portion of the entire Universe. We are no more than an expanding bubble of space that grew from a region smaller than a proton in one corner of the Universe. Elsewhere, forever inaccessible to us, may be an infinity of other expanding bubbles spread throughout space like the froth on a great sea.

15

THE GOLDEN AGE OF COSMOLOGY

Life after COBE

The most astonishing thing about COBE's discovery of hot spots in the microwave background was how readily most scientists came to believe in the result. "It was a terribly difficult measurement and the COBE team made only a marginal detection," says Jim Peebles.

So were the hot spots real? In the past, experimenters who had tried to measure the coldest thing in the Universe had mistakenly measured stray radiation coming from the Earth or the Galaxy, from their own equipment, the exhausts of their rockets and countless other spurious sources. So were COBE's hot spots really imprinted on the radiation from the beginning of time, or had the COBE team been hoodwinked by something altogether more mundane and closer to home?

"I still wake up in the middle of the night thinking 'Have we accounted for everything?' " says Dave Wilkinson. "That's my biggest worry about COBE—that we didn't measure everything we should."

One thing Wilkinson worries about is that stray radiation from the Earth may have got into the sensitive instruments by bending around the satellite's ground shield. The COBE scientists were unable to measure this effect, and instead had to estimate it from theory. Their calculations assumed, for the sake of simplicity, that the metal shield had a knife-sharp edge, but as Wilkinson points out the edge was bound to be ragged if looked at closely. The unanswered question is whether by oversimplifying the calculation, the team underestimated the amount of Earth radiation the COBE instruments were seeing.

Charles Bennett had another worry. He was concerned about spurious radiation coming from our Galaxy, and whether the team had subtracted it from their signal correctly. "I dedicated myself to satisfying myself that that wasn't what we were seeing," says Bennett.

The radiation given out by the Milky Way is complicated. It comes from glowing dust and also from electrons broadcasting radio waves as they spiral around the Galaxy's magnetic field lines.[1] The team had to have a theoretical "model" of how the different types of radiation should vary with wavelength, and then they had to make sure that their own measurements agreed with it. These were made at a mere three wavelengths, hence Bennett's worry. But in the end he satisfied himself that all was okay. "Whatever model you pick for the Milky Way, it accounts for virtually none of our signal," says Bennett. "That's what really convinced me we weren't seeing the glow of the Galaxy."

Of course, the COBE team could still have been hoodwinked. "If nature was malicious, it could have filled the halo of the Galaxy with three-degree dust," says Bennett. "That would look exactly like the cosmic background radiation!"

But many scientists not on the COBE team were ready to accept the result as correct. "I'm betting the COBE result is real and correctly measured," says Peebles. "First, I observe the people on the project—like Dave Wilkinson—and I have a deep faith in their ability to track down every last thing."

Peebles also believes that the evidence presented by the team is good. "Of course, it's statistical only—they don't have a picture of the face of God—but there are statistical tests of what they have found, and the ones I've examined seem pretty good.

"I'm betting it's good," says Peebles. "I would give you odds—not a million to one—but, oh, three to one, something like that."

By an odd coincidence, the hot spots detected by COBE were lurking just below the level that ground-based experiments could

[1]These are rather like the magnetic field lines revealed when iron filings are sprinkled about a bar magnet.

pick out. "We certainly lucked out finding them," says Dave Wilkinson.

"If COBE had been launched a year or two later, it would have been scooped," says Peebles.

It was clear that ground-based experiments, equipped with the latest detectors, would soon be able to see whether the satellite's hot spots really existed. "There was a minor gold rush to be the first person to check the COBE result from the ground," says Peebles.

"The unfortunate thing will be if people don't find anything," says Bennett. "Everyone will wonder which experiment was right, which was wrong. It could drag on forever."

"I'm pretty confident about what we did," says Bennett. "But if we made a mistake, I'd rather it was caught sooner than later."

As it happened, Bennett did not have to wait long.

Balloon confirmation

In December 1992, a team of astronomers from Princeton University, MIT and Goddard found hot spots in the cosmic background radiation. The hot spots were similar in all ways to those the COBE team had announced finding eight months earlier.

Ironically, the scientists involved—Lyman Page, Stephan Meyer and Ed Cheng—found their hot spots before COBE did. But the effect was so small that it took the team nearly three years to confirm that what they were seeing was really in the Big Bang radiation and not something else—for instance, a spurious signal in their instrument.

COBE had the huge advantage in that it orbited high above the atmosphere, which strongly absorbs the cosmic background radiation. The research budget of Page and his colleagues, on the other hand, did not quite stretch to COBE's $60 million. They had to get their peek at the Universe by hoisting their instrument package to high altitude beneath a balloon.

The balloon experiment used bolometer detectors that were twenty-five times as sensitive as those on COBE, the technology of which was frozen in the early 1980s. This meant that it could make

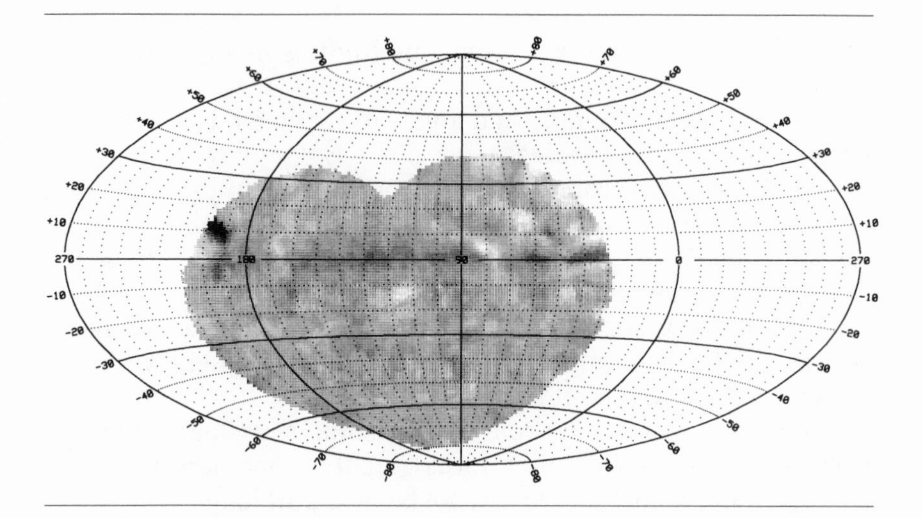

Confirmation of cosmic ripples. In December 1992, a team from Princeton, MIT *and Goddard announced that their balloon experiment had found hot spots in the cosmic background radiation. The hot spots were similar to those that* COBE *had announced finding eight months earlier. The dark spot on the left side of the image is actually the planet Jupiter.*

the same measurement as the satellite but 625 times faster. "We could do in six hours what COBE took a year to do," says Page.

Wilkinson admits that the detectors used on COBE were "medieval." But COBE succeeded, despite its insensitivity, because it was relentless, sitting up in orbit observing the background radiation day in and day out for more than a year. Instruments flown on balloons rarely get to observe for more than ten hours before high-altitude winds blow them out of range over the sea or over mountains.

Page and his colleagues had begun building their experiment in 1984. But when they first flew it, in 1988, disaster struck. "Our balloon burst!" says Page. However, in October 1989 the team took its instrument to the launch site at Fort Sumner in New Mexico. "This time everything worked perfectly," says Page.

The balloon reached an altitude of 25 miles, where it stayed for ten hours. During this time, the sensitive instruments on board observed the Big Bang radiation for a total of six hours, scanning a quarter of the entire sky for tiny temperature differences.

"When the data came down, it was clear there was a temperature variation in it," says Page. "The trouble was we couldn't tell whether the variation was in the background radiation or whether it had a more local source." It could have been caused by our Galaxy, the atmosphere or even by the instrument itself.

"One by one we eliminated all the possibilities," says Page. The instrument observed at four wavelengths—one more than COBE—and this helped Ken Ganga at Princeton determine the galactic emission, mainly from dust, and so subtract it from their signal.

Finally, the team had eliminated everything they could think of. They presented their result at a workshop held at Berkeley in December 1992.

The hot spots they had found were only 14.5 millionths of a degree hotter than the average temperature of the sky, slightly less than the 17 millionths of a degree found by COBE. But as in the COBE map of the sky the hot spots existed in all sizes from seven times the apparent diameter of the Moon up to a quarter of the sky.

Page and his colleagues compared their map of the sky with the one obtained by COBE to see if the bumps and the wiggles were the same. They were. "It's a pretty neat result," says Page. "We're really happy with it."

Meyer and Cheng had both worked on COBE as well. But several others working on the satellite made the same comparison independently, and they had confirmed that the agreement was very good. "The COBE team loves it," says Page.

What gave everyone so much confidence in the new result was that it had been obtained with an instrument that was very different from the one on board COBE. The balloon experiment used a single horn which pointed at forty-five degrees from the vertical and spun around a vertical axis once every minute. In this way, it was able to compare the temperature around a ring of sky. In the six hours the balloon's instruments were working, the Earth was turning beneath the sky, and so the horn was able to sweep out overlapping rings covering a quarter of the sky.

Apart from being a very different instrument, Page's experiment operated at slightly shorter wavelengths than COBE—between 0.44 millimeters and 1.8 millimeters. The difference was important since the signal from the cosmic background radiation is the residue left when the emission from our Galaxy is subtracted. At the shorter wavelengths of the balloon experiment, the main emission from the Galaxy was from warm dust, whereas the galactic emission seen by COBE was from electrons spiralling around magnetic field lines.

The galactic emission is the major uncertainty in any experiment, so it was a great relief when, after using two entirely different models—one for emission from dust and one for emission from electrons—the balloon and space experiments were left with precisely the same hot spots in the cosmic background radiation.

So the COBE result was vindicated.

"Back in 1974," says John Mather, "we set out to do a job so terrific that no matter what the theorists came up with we'd have all the data that anyone could get. We would reach the limits set by our location in the Universe. You can't send a space probe out of our Galaxy. You can't even send one out of the Solar System. But we said we would do the best we can, living here. We've just about done that."

The small scale

With both of COBE's major discoveries now confirmed, attention turned to studying how the Big Bang radiation varied across the sky on an even finer scale. Even the smallest lumps of matter COBE had seen in the early Universe were bigger than the largest collections of galaxies astronomers have so far seen in today's Universe. But lumps of matter in the early Universe should have arisen in all sizes. So if people zoomed in on small portions of the sky, they ought to be able to see lumps small enough to have been the seeds of individual galaxies like the Milky Way.

The smart money was on Phil Lubin of the University of California at Santa Barbara being the first to find such lumps. Lubin's team had set up an experiment at the South Pole, where the air was free of troublesome water vapor. The aim was to discover hot spots in the sky as small as the apparent diameter of the Moon—fourteen times smaller than the smallest spotted by COBE.

Lumps of matter of this size were potentially much more important than those found by COBE. Those lumps were larger from end to end than any light signal could have traversed 300,000 years after the Big Bang, so there was no way they could have been affected by any processes occurring in the Universe at that time. If they told scientists about anything, it was of a much earlier epoch, perhaps the first split second of the Universe. However, lumps the size of those sought by Lubin's team were small enough to have been affected by processes occurring 300,000 years after the Big Bang. Potentially, they would provide a panoramic window onto the Universe at the instant galaxy formation got under way.

If anything, the hot spots on the small scale should be hotter than those found by COBE. The reason is that COBE's hot spots were not caused directly by matter but indirectly, through its gravitational effect on the fireball photons. But on the small scale, theorists fully expected to see the direct effect of electrons on the photons of the fireball radiation. Before they combined with protons to form atoms, these electrons could have collided with photons, boosting their energy and making them appear hotter. At present, no such effect has been seen by Lubin or other astronomers, so the theorists are a little worried.

Other astronomers have been looking for even finer temperature variations in the Big Bang radiation. In fact, Bruce Partridge, who is currently at Haverford College, along with Ed Fomalont and Ken Kellerman have been carrying out such experiments for years using the Very Large Array of radio telescopes in Socorro, New Mexico. This instrument consists of twenty-seven dish antennas arranged out in the desert along the arms of a Y-shaped railway track. Since the degree of detail a telescope can see depends on how large it is, the Very Large Array—which can be as much as twenty miles across—is able to see very fine detail indeed in the sky.

But one of the big drawbacks of such an instrument is that other people want to use it. So Partridge has rarely received more than a week or so of continuous time to use the Very Large Array, although, as COBE showed, having a lot of time to observe the cosmic background radiation is the key to success. Another drawback of the Very Large Array is that it is not specifically designed for cosmic background experiments but for more general radio astronomy. What is really needed to pursue this line of work is a permanent array of properly designed radio telescopes.

Such an array is planned at the University of Cambridge. Anthony Lasenby and his colleagues want to build an array of between fifteen and twenty microwave horns, each no more than a meter or so in length. The horns would be fixed to a circular turntable and would be able to look for hot spots as small as half the size of the Moon.

The astronomical community is currently waiting to see whether Lasenby's ambitious project will get the funding it needs to proceed.

The Golden Age of Cosmology

In the wake of COBE, there is now intense interest in the afterglow of creation. We now know that written across the sky is the story of the early Universe, and we are only just beginning to read that story.

With the help of COBE, we have seen the Universe through the most sensitive microwave glasses ever made. What at first appeared to be the unbroken whiteness of the fireball has resolved itself into a complex patchwork of light and shadow, telling us of the birth of the giant clusters of galaxies at the beginning of time.

Heartened by COBE's discovery, an army of people equipped with microwave glasses is now peering ever more closely at Big Bang radiation. They are zooming in on smaller and smaller patches of sky in the hope of finding the seeds of individual galaxies like the Milky Way.

Until now, we have managed to glean only a few scraps of fundamental information about the nature of the Universe. But the cosmic background radiation promises to greatly increase that knowledge. The temperature of the fireball radiation is already the

The biggest eye in the world. The 10-meter Keck telescope stands on the 14,000-foot-high summit of Mauna Kea on the Big Island of Hawaii. Its "fly's eye" is made of 36 hexagonal segments. A computer, which controls the movement of each segment, keeps the surface in shape as the giant mirror swings around the sky.

most precise thing we know about the Universe and we are only just beginning to decode the secrets of this oldest fossil in creation.

This is the Golden Age of Cosmology. As we begin to read the message written on the microwave sky, a new generation of big telescopes, like the 10-meter diameter Keck telescope in Hawaii, is coming on stream. Equipped with hypersensitive detectors, such telescopes will be able to see galaxies at enormous distances when they were young. "Very soon, there's going to be an explosion in our understanding of the early Universe," says Jim Peebles.

The Russians are said to be considering another satellite to observe the fireball radiation, a follow-up to their Relict I. And at the Goddard Space Flight Center there is even talk of building a COBE II to measure the spectrum of the Big Bang radiation at the longer wavelengths that COBE could not reach. It seems that the afterglow of creation has caught everyone's imagination. "There's an awful lot more life left in this beast," says John Mather.

The future of the background radiation

It has been a long slog—wringing the precious secrets from the cosmic background radiation—but we have been lucky. Though the afterglow of creation is terribly faint, it is still possible to pick it out from the bright microwave glow of our Galaxy. Had humans evolved much later in the history of the Universe, it might have been a different story . . .

In another 15 billion years, the remorseless expansion of the Universe will have driven the galaxies twice as far apart as they are today.[2] The photons of the background radiation will be stretched to longer wavelengths and diluted even more. The Universe, instead of being pervaded by a background glow at a temperature of 3 degrees, will be filled with radiation at only 1.5 degrees above absolute zero. From inside a galaxy like the Milky Way it will be hard to pick out ripples like those found by COBE. Hard but not impossible. It will simply take a lot more patient observing.

But when the Universe is three times the age it is now, the temperature of the cosmic background radiation will be only a third of what it is today; when the Universe is four times as old, just a quarter. By the time 150 billion years have elapsed since the Big Bang, the relic of the fireball will have all but died out. It will be a mere 0.3 degrees above absolute zero. If there are any intelligent species around 150 billion years from now, they will not be nearly as fortunate as we have been. In their Universe, the afterglow of creation will be essentially undetectable, its secrets forever beyond reach.

The fate of the fireball radiation in the very distant future depends on whether the Universe contains enough matter—both visible and invisible—to one day halt and reverse its expansion. If it does not have enough matter, and the Universe is destined to expand forever, its dying galaxies becoming ever more isolated is-

[2]To make the numbers simple, I have ignored the fact that the combined gravity of all the matter in the Universe is braking its expansion.

lands in an ever-growing ocean of space, then the radiation will simply be diluted out of existence.

On the other hand, if there is enough mass in the Universe to halt its expansion and turn it into a runaway collapse, the relic radiation will be rescued from such an ignominious end. As the Universe shrinks inexorably down to a Big Crunch—a sort of mirror image of the Big Bang, in which all of creation is squeezed again into an impossibly small volume—the background radiation will get hotter and hotter as it is squeezed to shorter and shorter wavelengths. No longer a few degrees at radio wavelengths, it will appear a few tens of degrees in the infrared. Then, as the burned-out hulks of galaxies are crushed together, the Universe will blaze again with visible light, corresponding to a temperature of thousands of degrees.

This will be the mirror image of the epoch probed by COBE. Atoms, instead of forming for the first time, will be broken apart. The Universe, instead of becoming transparent to radiation, will become utterly opaque. The billions of years of domination by matter will be over and radiation will at last be king again.

In the last minutes before the Big Crunch, all of creation will be a raging inferno. The ferocious light of the fireball radiation will begin to blast apart the nuclei of atoms into their constituent protons and neutrons. Soon all traces of ordinary matter will be expunged forever from the Universe.

The fireball radiation will have returned whence it came. No longer the afterglow of creation, it will now have transformed itself into the deadly aura of destruction.

EPILOG

Wrinkles in Time

Although the Cosmic Background Explorer was a resounding success, exceeding the expectations of even the most extreme optimists at NASA, one thing had taken some of the shine off the success: the bad feeling created by the actions of George Smoot. However, Smoot was writing a popular book about the quest for cosmic ripples with Keay Davidson, a respected newspaper journalist on *The San Francisco Examiner.* It was a golden opportunity for Smoot to smooth things over with the COBE team . . .

However, things did not get off to an ideal start when the COBE team first found out about the upcoming book in a stray reference in the media. "George did not tell us himself," says John Mather. "We actually read it in a column in *Scientific American.*"

"I believe they learnt about it the same time as I did," Smoot says, who remembers things differently.

Whatever the truth, the prospect of an account by Smoot caused concern among some members of the team. "In normal circumstances, nobody would be worried if someone was writing a book," says Bennett. "However, because of the way COBE had become synonymous with George's name there was a definite concern that his contribution might get emphasized at the expense of the team."

Such concerns were voiced to Smoot by some members of the COBE team, including Rai Weiss and John Mather. "I insisted that we get a look at the draft of the book," says Mather. "After all, we had a written agreement that we would all get to see what any of us was writing about the project."

Smoot readily agreed and Keay Davidson showed the team an early draft. "Everyone who asked to see it, saw it," says Bennett. "I chose not to comment, but team members such as John Mather took the matter very seriously and sent detailed suggestions."

"There were no major things wrong with George's account," says Mather. "It was much more a matter of omissions. For instance, there was hardly a mention of Bennett and the others who had worked with Smoot on the DMR smoothness experiment. That was a little unfair because Chuck Bennett was the man on the spot every day to make the DMR happen," says Mather.

Smoot insists that Bennett is in the book. "Some people might even say I gave him too much credit," he says.

According to Wilkinson, the book was much more of a personal perspective than anyone had expected. "It was the flavor of the thing that was a bad idea," he says. "I definitely felt that George had diminished the contributions of other people. I told George it would only exacerbate problems with the team."

One man who was particularly upset by the account was a Goddard engineer named Tony Fragomeni. As the "observatory manager," Fragomeni had been responsible for designing and building the spacecraft to accommodate the scientific instruments. "I was very unhappy about the general tone of the excerpts I saw," says Fragomeni. "It was too much of Smoot doing the whole thing himself. The mission originated with John Mather."

Fragomeni says he "got disturbed" and faxed Keay Davidson about changes he thought should be made. When he received no reply, he became anxious that his suggestions might not be incorporated and got a lawyer to draft a letter. "I insisted that my name be removed from the book," says Fragomeni. "I simply didn't want to be associated with it."

Fragomeni's wishes were respected. "Things attributed to me are still in the book," says Fragomeni. "But my name has been taken away."

Gloomy predictions borne out

Smoot and Davidson's book was published under the wonderfully imaginative title of *Wrinkles in Time*. On the cover, the publisher,

Little Brown, reproduced Stephen Hawking's extravagant quotation about the cosmic ripples discovery being the scientific discovery of the century—if not of all time. It was a brilliant move. Hawking was not only the most famous scientist on the planet but he had also written the biggest selling popular science book of all time.

When the members of the COBE team saw copies of *Wrinkles in Time* they were pleased that changes had been made since the early draft. "In most cases, George did something in response to the comments of the team," says Bennett.

However, Wilkinson's gloomy prediction that the book would exacerbate problems with the COBE team was borne out. "Everything was told in an incomplete way to make George look good," says Mather.

"I certainly didn't try to make myself look bad," admits Smoot. "But then again I didn't go out of my way to make myself look good either."

As for crediting his colleagues, Smoot admits he did not indicate exactly how much work each person did. "If you start trying to say what everybody did and say how important they were relative to everybody else, it's an endless pit," he says. "I tried to do it. But, in the end, the editors cut a tremendous amount of stuff out."

Smoot maintains he did everything possible to share credit. "I went out of my way to name the people and their contributions—including putting a list of everyone who ever worked on COBE at the back of the book."

However, far from satisfying Smoot's colleagues, the list, which included about 1500 names and had been sent by John Mather, actually inflamed them. "The publication of the whole list allowed George to say he had given everyone due credit," says Bennett. "However, the only thing the list conveys to a reader is: these are the people who helped me."

"A more accurate description would be: these are the people who made the discovery," says Mather.

"Let's face it, the list was a futile attempt to share credit," says Wilkinson.

"The list mixed together scientists and technicians and anyone who contributed even a tiny bit to the satellite," says Bennett.

"No distinction was made between those with a major and a minor role in COBE."

However, it has to be said that not everyone was unhappy with the publication of the long list of names. Quite the opposite. "Some of the programmers and engineers on COBE were very happy indeed," says Bennett. "They thought it was great seeing their names in print."

A positive response

Outside the COBE team the reaction to *Wrinkles in Time* was largely positive. "Considering all the hype there had been, it was better than I expected," says Martin Rees of the University of Cambridge. "It's not a brilliant book but it's not a terrible book either."

Those who reviewed the book in newspapers and magazines were similarly upbeat. Smoot's co-author Keay Davidson was very pleased. "I was absolutely taken aback by the reaction," he says.

However, few of the people who reviewed the book were in a position to judge whether it was a fair account. One exception, however, was Joel Primack, a cosmologist at the University of California at Santa Cruz, who was asked to write a review by the magazine *Physics Today*.

Primack's eyebrows were raised when he read of the proposal to remove the so-called quadrupole component of the DMR's smoothness measurement. The removal of the quadrupole signal was in many respects analogous to the subtraction of the dipole signal, caused by the motion of the Earth through the background radiation. The big difference was that the quadrupole signal was due largely to the microwave glow of our Galaxy. Its removal from the DMR data would therefore ensure that any leftover variation in sky brightness was due to "ripples" in the Big Bang radiation itself.

This was all familiar territory to Primack. But what grabbed his attention was the claim by Smoot that, in February 1992, it was he who had come up with the idea of subtracting the quadrupole signal. Primack was sure that in fact Ned Wright had done such a calculation, and that Wright had done it well before late February 1992.

Primack called Wright at the University of California in Los Angeles. Wright confirmed that he had indeed carried out the quadrupole subtraction in November 1991. He even sent Primack documentary evidence. According to Wright, it was actually Steve Meyer, another cosmic background scientist, who had first proposed the idea as early as September or October 1991.

Smoot does not dispute that Meyer came up with the idea of subtracting the quadrupole signal. However, he says Meyer's idea was to subtract it from data he and his colleagues had obtained from a balloon experiment to look for ripples in the Big Bang radiation. This was specifically so that the data, which covered only a part of the sky, could be compared with the whole-sky data from the DMR. "It wasn't a clever idea to analyze the DMR data but a way to *compare* the data from Meyer's experiment with the DMR data," says Smoot. "There's a difference."

Primack also phoned Wilkinson and Weiss. He learnt that all three were upset with Smoot's failure to give proper credit, especially to junior members of the team. The upshot of all this was that, when Primack submitted his review to *Physics Today* at the end of January 1994, it contained a mild paragraph saying the COBE team was upset.

What happened next surprised Primack. "*Physics Today* sat on the review for most of 1994," says Primack. "When I telephoned the editorial director, Gloria Lubkin, to find out what was happening, she told me the magazine was afraid of a lawsuit."

When Primack reread what he had written, he shook his head with disbelief. "My review was definitely a favorable one," he says. "I actually liked the book."

However, this was not the end of the affair. "Later, *Physics Today* tried to get me to remove the paragraph about the COBE team being upset," says Primack. When Lubkin was later asked about the delay in publishing Primack's review, she refused to discuss the matter.

"I stuck to my guns," says Primack. "It was basically a good review, after all." Confirmation of this came in October 1994 when the unexpurgated review finally appeared. "I got a nice postcard from Keay Davidson," says Primack. "He said he loved the review."

Smoot, however, hated the review. "I didn't think it was appropriate for a book review to say that some unnamed people on the COBE team were unhappy with the credit," says Smoot. "If people are going to criticize me, they should put their names in. When I read the review, I took it as a personal affront."

An opportunity missed

Wrinkles in Time had provided George Smoot with a golden opportunity to smooth things over with the COBE team and dispel any bad feelings. However, the view of some members of the COBE team was that Smoot had not taken that opportunity. "George really missed a good chance," says Mather.

"For all those years the team functioned beautifully," says Wilkinson. "Everyone was so pleased at how well it went. To have it come apart at the last minute was very sad."

"If ever there was a project done by an integrated team it was COBE," says Fragomeni. "At Vandenberg, before the launch, we had about eighty people at any one time, working up to eighteen hours a day, three to four days at a stretch. I've worked on big engineering projects for forty years and there is a tendency for flare-ups. But during the seven weeks we were at Vandenberg I don't recall a single flare-up."

But if there was sadness it was not only sadness that the "perfect team" had been damaged but sadness that the person responsible was a long-time friend of many of those on the COBE team. "I feel sorry for George," says Dave Wilkinson. "I've always liked him. But he has hurt himself over this. His reputation among scientists has suffered."

"I certainly think that when people say stuff like this about you it hurts," says Smoot. "It hurts your reputation. It makes people wonder about what's going on. It also hurts COBE, which is a wonderful achievement by a large number of people."

However, Smoot thinks it will all blow over in the long run. "People are getting upset about this but it's crazy," says Smoot. "I think if you do a literature search and look under COBE and my name, you'll find it's a hundred to one. In the end, it won't be my name or anyone else's name that's remembered—it'll be COBE."

INDEX